T0140661

Evolutionary Approach to Machine Learning and Deep Neural Networks

Hitoshi Iba

Evolutionary Approach to Machine Learning and Deep Neural Networks

Neuro-Evolution and Gene Regulatory Networks

Hitoshi Iba
The University of Tokyo
Tokyo
Japan

ISBN 978-981-13-4358-2 ISBN 978-981-13-0200-8 (eBook)
https://doi.org/10.1007/978-981-13-0200-8

Softcover re-print of the Hardcover 1st edition 2018

Printed on acid-free paper

This Springer imprint is published by the registered company Springer Nature Singapore Pte Ltd.
part of Springer Nature
The registered company address is: 152 Beach Road, #21-01/04 Gateway East, Singapore 189721, Singapore

Preface

Around 1990, I learned about GP (Genetic Programming) and developed interest in this area shortly after learning about GAs (Genetic Algorithms). At that time, the term GP was not yet established, and the concept was denoted by terms such as structural GA and tree representation GA [1–3]. Since I had been researching AI so far, I presumed that the only goal of GA was optimization (my opinion has changed since), and I considered GA to be somewhat unsatisfactory. Consequently, as I assumed that GA could not be used for handling knowledge representation, programs, concept trees, and similar notions, I attempted to extend it. At exactly the same time, when I presented my research to Dr. Philip D. Laird from NASA, who was a Visiting Researcher at the Electrotechnical Laboratory. Dr. Laird is a researcher in machine learning and is renowned for his book *Learning from Good and Bad Data* [4]. He introduced me to the research of Prof. John Koza of Stanford University. I still remember the excitement I felt while reading the technical reports [5] written by Koza over the new year holidays. These reports were later compiled in a massive volume that exceeded 800 pages [6]. Afterward, I stayed at Stanford University in his laboratory and had a wonderful research life there.

Recently, an increasing number of researchers and students are specializing in only GA and GP (and fields centered on them). Also, in comparison with the time when AI and GA/GP were created, there are fewer and fewer researchers with unique personalities. Although this is not in anyway disappointing, it cannot be understood unless one has experienced the turmoil and the emotions associated with the establishment of a new field. I expect such pleasant nervousness and excitement to be maintained in this academic society as well, setting the path for the development of a healthy community. This is one of my motivations in writing this book.

The characteristics of EA (evolutionary algorithm) can be summarized as follows:

- **Parallelism**: A large number of individuals can be searched simultaneously as a group. This is suited to advanced parallel application and can fully utilize computer power.
- **Searchability**: EA does not presume a deep knowledge of the search space (calculation of differentiability, gradient, etc.)
- **Diversity**: As there are a wide variety of individuals within the group, it excels in adapting to environments with dynamic changing problems and noise, and the solutions obtained have a high level of robustness.

We can introduce such promising biological knowledge, e.g., symbiosis, coevolution, and habitat isolation, into the calculation mechanism in EA.

For this purpose, this book provides theoretical and practical knowledge about a methodology for EA-based search strategy with the integration of several machine learning and deep learning techniques, e.g., memetic concepts, neural networks, Gröbner bases, belief networks, and affinity propagation. The development of such tools contributes to better optimizing methodologies.

The concepts presented in this book aim to promote and facilitate the effective research in EA approaches in both theory and practice. However, the contents of the book would be valuable to different classes of readers because it covers interdisciplinary research topics that encompass machine learning methodologies, deep neural networks, neuroevolution, and gene regulatory networks. EA practitioners will find this book useful for studying evolutionary search and optimization techniques in combination with deep learning and machine learning frameworks.

In addition, while still undeveloped and unrefined, some of the research examples shown in this book include contents worthy of scrutiny and publication in a thesis. However, it is not my wish to preemptively dismiss interesting and challenging attempts in actively developing fields such as AI and EA, so this book actively introduces such research topics. I sincerely hope readers further develop these research examples to achieve fruitful results.

In parallel to EA and GP, which constitute rather general search methods, researchers in this field must be equally versatile and must look beyond their own fields of specialization to learn about new topics in order to pursue new models and applications. As long as research is conducted with this attitude, it will be possible to "sustain the dream" for both a scientific community and individual researchers. I hope that this book will help readers make such an academic venture in EA and AI fields.

Tokyo, Japan
January 2018

Hitoshi Iba

References

1. Iba, H.: On Information Evolution and Creative Learning. *JSAI Fundamentals Workshop*, SIG-F/H/K-9001-10 (1990)
2. Iba, H.: Adaptive Learning of Structural Expressions and Its Application. *AI Workshop of the Information Processing Society*, AI-76-2, JSAI (1991)
3. Iba, H.: Learning of High-level Knowledge based on Adaptive Method. *Fifth Annual Conference of JSAI* (1991)
4. Laird, P.D.: Learning from good and bad data. Springer (1988)
5. Koza, J.: Genetic Programming: A Paradigm for Genetically Breeding Populations of Computer Programs to Solve Problems. Report No. STAN-CS-90-1314, Department of Computer Science, Stanford University (1990)
6. Koza, J.: Genetic programming, on the programming of computers by means of natural selection. MIT Press (1992)

Acknowledgements

To all those wonderful people, I owe a deep sense of gratitude, especially now that this book project has been completed. Especially, I acknowledge the pleasant research atmosphere created by colleagues and students from the research laboratory associated with Graduate School of Frontier Sciences and Information Science and Technology at the University of Tokyo.

I am grateful to my previous group at Electrotechnical Laboratory (ETL), where I worked for ten years, and to my current colleagues at Graduate School of Engineering of the University of Tokyo. I wish to express my gratitude to SmartRams Co., Ltd. for cover image design.

And last, but not least, I would like to thank my wife Yumiko and my sons and daughter Kohki, Hirono, and Hiroto, for their patience and assistance.

Tokyo, Japan
January 2018

Hitoshi Iba

Contents

Chapter 1
Introduction

When you were a tadpole and I was a fish In the Paleozoic time,
....

(Evolution, by Langdon Smith: 1858–1908)

Abstract This chapter gives a basic introduction to evolutionary mechanisms and computation. We explain a fundamental theory of evolution and some debatable issues, such as how complex facilities like eyes have evolved and how to choose next generation from elite members. Thereafter, the method of evolutionary computation is described in details, followed by GP frameworks with several implementation schemes.

Keywords Evolutionary mechanisms · Genetic algorithms (GA) · Evolutionary computation (EC) · Genetic programming (GP) · Linear genetic programming Cartesian genetic programming (CGP) · Interactive evolutionary computation (IEC)

1.1 Evolution at Work

In Western Australia, there is a World Heritage Site known as Shark Bay. On the coast here, Shell Beach is famous for being full of small white clams. The snow-like white beach stretches out for a total of 120 km, and the contrast with the blue sea makes it truly beautiful, and definitely worth seeing. On the other hand, in Hamelin Pool in Shark Bay, you can see the world's oldest organism, the stromatolite, which has existed since 2.7 billion years ago. The stromatolites are cyanobacteria[1] fossils and are a rock with a layered structure. Even today, this rock continues its slow growth of approximately 0.3 mm per year. In the Precambrian period, they existed

[1]Cyanobacteria are blue-green algae, and when it is in its collected state on the surface of water, it is known as green laver. It can be observed when the sun's rays hit the green laver, and large numbers of small oxygen bubbles in the surrounding area.

H. Iba, *Evolutionary Approach to Machine Learning and Deep Neural Networks*, https://doi.org/10.1007/978-981-13-0200-8_1

(a) The Hamelin Pool Stromatolites, located in Shark Bay on the Coral Coast of Western Australia.

(b) Stromatolite fossils, in which produced gas bubbles can be found.

Fig. 1.1 Stromatolite

in all areas of the world, but they now only exist in two locations on the earth.[2] The stromatolite rocks, countless in number and perfectly clear off the cove, are truly impressive (Fig. 1.1). As I walked along the seashore, it may have been my mind playing tricks on me, but it felt as though the salt water was hurting my feet. The sea water in the Hamelin Pool has approximately twice the normal levels of salt content, and seaweed does not grow there. For this reason, predatory creatures cannot evolve there, and the stromatolites have been able to survive.

In actual fact, this small organism and its mass can be said to exhibit the mobility of evolution. Cyanobacteria have a chloroplast known as chlorophyll a and use the sun's rays to perform photosynthesis. In other words, organic materials are synthesized from carbon dioxide and water, and oxygen is discharged. The reason that a huge quantity of oxygen was emitted in the earth from 2.2 billion to 1.9 billion years ago was because oxygen was waste matter during photosynthesis.

In evolutionary terms, oxygen, which had been rare until that time, began to increase, and the elements of life underwent a change. That is to say, oxygen has high activity and vitality; however, on the other hand, oxygen has high oxidative capacity and is toxic. Therefore, we saw the appearance of aerobic bacteria, which had resistance to oxygen and could be used. It is thought that, starting from there, evolution accelerated explosively. In other words, the oxygen bubbles of stromatolites are the motivating force of evolution. In one sense, if this bacterium was not around, there might not have been any organisms around, including ourselves. Put another way, everything may have started from this point. It feels as though, in this small bacterium producing bubbles, we saw the source of creative phenomena exceeding human knowledge. The evolution of organisms is neither inevitable nor coincidental.

Let us consider, therefore, what is necessary for things to evolve. Charles Darwin acknowledged that groups of organisms have the following characteristics.

---

[2]They exist in Cuatro Cienegas in Mexico as well as in Shark Bay.

(a) Giant tortoises. (b) Land iguana.

(c) Sea iguana. (d) Hybrid species.

Fig. 1.2 Animals in Islas Encantada (Galapagos Islands)

- Groups of organisms all produce descendants in an explosive manner.
- Transformation is, in fact, a characteristic held by all animal and plant groups.
- Those organisms that best adapt to their environment survive.

Today, transformation is known as genetic recombination (crossover or mutation). Furthermore, the first characteristic of "producing descendants in an explosive manner" is based on the principle that "if population is not restricted, it will grow exponentially, but life resources can only grow on an arithmetic progression" as developed in "Essays on the Principle of Population" by Thomas Robert Malthus.[3] It is thought that Darwin arrived at the theory of evolution by reading "Essays on the Principle of Population." Darwin stated that if adaptive characteristics, to any extent, are inherited from these three characteristics, these advantageous traits are communicated to the next generation.

The Galapagos Islands are famous as the place where Charles Darwin stopped while traveling on the British Navy survey ship, the Beagle, and at that time, noticed that the shape of the finch bird's beak differed slightly depending on the type, and that the shell of the tortoise differed depending on the island, leading him to come up with the theory of evolution (Fig. 1.2a). However, this story is somewhat exaggerated.

[3]English economist (1766–1834). He explains population growth to be the ultimate cause of poverty and crime in his essays.

Darwin had not deeply researched this bird locally. If you read his "Origin of Species," you will see that the Galapagos data is not used so much, and that it is no more than a discussion centering on artificial selection.

However, even today, this island attracts intense scrutiny as a center for evolutionary research. For example, Peter Grant et al. from Princeton University have thoroughly studied the finch for over 20 years and have discovered that selection pressures resulting from feed yields, as a result of climate change, have dramatically changed the beaks of the finches. Furthermore, this evolution has not occurred over the normally considered timescales of several tens of thousands of years, but has been seen to occur at an incredibly fast speed (on the order of several years) [6, 16].

Additionally, we have learned that two species of the iguana, the land iguana and sea iguana, exist in the Galapagos Islands. The yellow iguana was originally a land animal and ate cholla buds (Fig. 1.2b). However, due to the coevolution of cholla, this food source has become insufficient. Coevolution is when two or more species evolve while influencing each other. The result of coevolution is evolution in the order of (1) competition (arms races), (2) parasites (commensalism), (3) coexistence (collaboration). In this situation, the cholla, land iguana, and giant turtle became embroiled in an arms race.[4] The cholla grew tall so its bud section would not be eaten, and developed thorns in their stems. The land iguana cannot climb the trunk of the cholla. In this way, food became insufficient as part of coevolution with the cholla. The result of this was that the sea iguana, which could swim in the sea and eat the seaweed attached to the rocks, has evolved. The sea iguana has a completely different ecology than the land iguana. Its body color is black to accumulate sunlight, and in order for it to hold its breath, dive in the water, and return to the land, it is able to expel salt water from its nose (Fig. 1.2c). Interestingly, recently there are stories of hybrid iguana being born as children of male sea iguanas and female land iguanas. These are pink iguana, and they have both sets of characteristics (Fig. 1.2d). They have nails and are able to climb the trunks of the cholla, but at the present time they are not able to proliferate.

It is the objective of the evolutionary method to realize a calculation system (evolved system) based on this method of thinking. The thinking behind evolution not only concerns nature but reaches all areas. For example, let us consider one major invention of humankind, the airplane. The Wright Brothers, who invented it, succeeded in the first takeoff of the Wright Flyer after repeated trial and error. While repeatedly failing during the development phase, they took good ideas from those in the past and attempted to perform minor changes (=mutations). Additionally, they combined good ideas with each other (=crossover) to create the next prototype. After repeated prototypes, they increased accuracy, and this is how evolving organisms work. Furthermore, even in the case of the proliferation of livestock and racehorses, which have been carried out since ancient times, humans have used evolutionary mechanisms unconsciously. Evolutionary computation uses the same mechanism.

In the following section, we explain the basic principles of evolutionary computation. However, before that, we shall explain a doubt which troubled Darwin and a successful case of evolution simulation.

---

[4]See p. 189 for the robotics application.

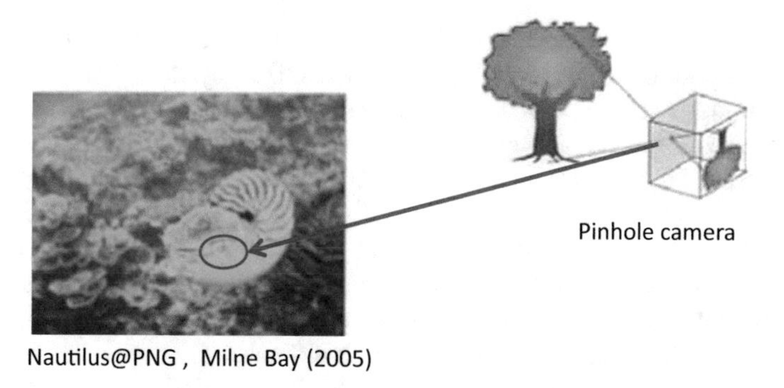

Fig. 1.3 What good is half an eye?

1.2 Have We Solved the Problem of the Evolution of the Eye, Which Troubled Darwin?

As we all know, the eyes of mammals, such as humans, are extremely complex organs. The eye is formed of a large number of systems, such as the retina, pupil, iris, cornea, lens, and the optic nerve. For example, the lens deflects light as it passes through the pupils, and the retina known as the fundus plays the role of a lens linking the image. The evolution of the eye troubled Darwin greatly in terms of the concept of evolution. In fact, in the "Origin of Species," it is stated that the evolution of complex functions such as eyes is considered to be "absurd in the highest degree" and further confessed that the eye made him shudder even later on.

When considering the theory of evolution in regard to the eye, the following two doubts appear [5, 13].

- Do the first steps and intermediate stages of evolution (related to the eye) remain?
- Do complex organs such as the eye actually undergo mutation?

People who criticize evolution also targeted this area.

The first doubt, "What good is half an eye?" is solved as one gets to know a variety of organic forms. For example, in the case of the cyanobacteria, which is a constituent element of stromatolites (see Fig. 1.1), they are known to be linked to light response, known as phototaxis. The nautilus (relative of the ammonite and in the same group as the argonaut living in the shell) has an eye like a pinhole camera (Fig. 1.3). The functions of these eyes are extremely primitive, but they have adapted well to the environment they inhabit. In other words, despite being simple eyes, they are better than nothing. The protein known as opsin that codes opsin genetics used in the field of optogenetics[5] is known for its photosensitivity. There is a theory that this gene appeared several hundred million years ago and created the oldest eye.

[5]This is technology that generates protein molecules activated by light genetically in special cells and operates special functions using light.

The second doubt remained a riddle for a long period of time. Certainly, simple eyes such as phototaxis and pinhole cameras seem like they could evolve from mutation; however, what about the eyes of mammals, which are like cameras with lenses? In order to reach this point, there are many blind alleys. The fact that there are still simple eyes like those of the nautilus and cyanobacteria existing today suggests that, beyond this point, an extremely difficult path opens up. Richard Dawkins[6] referred to this as "climbing an impossible mountain." He equates evolution with mountain climbing. Until the peak of the mountain, equating to the eyes of mammals, there are a large number of small peaks (simple eyes, such as those of nautiluses and cyanobacteria) that make climbing to the peak difficult.

In recent years, elucidation of this problem has been carried out using an evolution simulation [12, 13]. The method used here is the evolutionary computation explained in the next section. The simulation starts from a flat block of photosensitive cells sandwiched between a transparent protective layer and a black pigment layer (upper left of Fig. 1.4). This is the undeveloped stage of the eye, and the structure is like that of flat skin. Here, the evolution speed is 0.05% (change rate per generation). This is the minimum estimate, and the actual evolution speed may be faster. For example, it is known that the pigmentation of photoreceptors in crustacea changes much more rapidly than this value. Furthermore, the question of to what extent this eye performance (conformance) links to a clear image is measured and calculated optically.

When looking at the experiment results (Fig. 1.4), with step 2 and 3, the photosensitive organ layer and pigmentation layer (=retina) are ingrown to form a semi-circle. The protective layer becomes deep to form a vitreous body that fills the cavity. The refractive index of this vitreous body is set to 1.35. This value is slightly higher than the refractive index of water; however, this is not sufficient to cause a meaningful optical effect on the vitreous body. As a result, in steps 4 and 5, the retina continues to grow without changing the refraction diameter. Then, the hole in the retina is deepened in order to tighten the aperture of the peripheral section. In steps 6–8, the result of adding the refractive index locally is that the refractive index distributed lens appears. The lens central refraction rate expands from the initial rate of 1.35 to 1.52. At the same time, the form of the lens changes from an ellipsoid body to a spherical shape, and it transitions to the center of curvature of the retina. As the lens shrinks, and the opening widens, a flat iris gradually appears. The focal distance of the lens (f in Fig. 1.4) gradually shortens, and in step 8, focal distance f becomes equivalent to the distance (P) to the retina, and the freshness matches the focus. In the diagram, the diameter of the normalized receptor (relative change in diameter) is shown as d.

[6]English evolutionary biologist/animal behaviorist (1941–) who has written many general books and general introductions to biology and, as a result, espoused thinking on the "Selfish gene" (see p. 233), "Meme" (cultural information replicators), and "expanded phenotypes" (host operation due to the parasites, dams made by beavers, and mounds of white ants can be seen as phenotypes). His revolutionary ideas and provocative comments about evolution are still causing many discussions. He is a famous atheist.

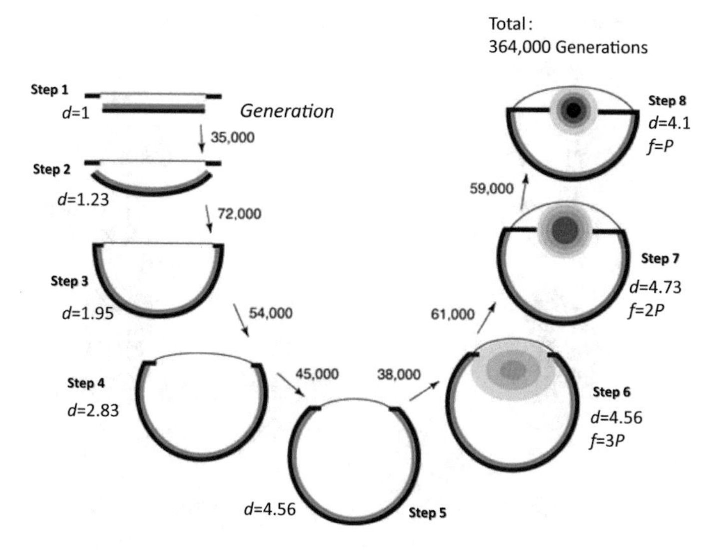

Fig. 1.4 Simulation of the eye's evolution (Reproduced from [12])

With this simulation, it did not take even 400,000 generations for the dense gradient type eyes such as those in fish appearing from the first undeveloped stages to evolve. In normal organisms, one generation is approximately one year, so this would be 500,000 years or less. This is no problem on an evolutionary scale. Furthermore, in the case of drosophila, as one generation is only 28 days, this would be 7,700 years. Complex eye evolution has been confirmed using this kind of computer simulation. Would this allow Darwin to sleep easily?

1.3 Evolutionary Algorithms: From Bullet Trains to Finance and Robots

Evolutionary computation is an engineering method that imitates the mechanism of evolution in organisms and applies this to the deforming, synthesis, and selection of data structures. Using this method, we aim to solve the problem of optimization and generate a beneficial structure. Common examples of this are the computational algorithms known as Genetic Algorithms (GA) and Genetic Programming (GP).

The basic data structures in evolutionary computation are based on knowledge of genetics. Hereafter, we shall provide an explanation of these.

The information used in evolutionary computation is formed from the two-layer structures of PTYPE and GTYPE. GTYPE (genotype, also called genetic codes, and equating to the chromosomes within the cells) are, in a genetic type analogy, a low-level, locally regulating set. This is the evolutionary computation to be operated on, as described later. The PTYPE is a phenotype and expresses the emergence of behavior

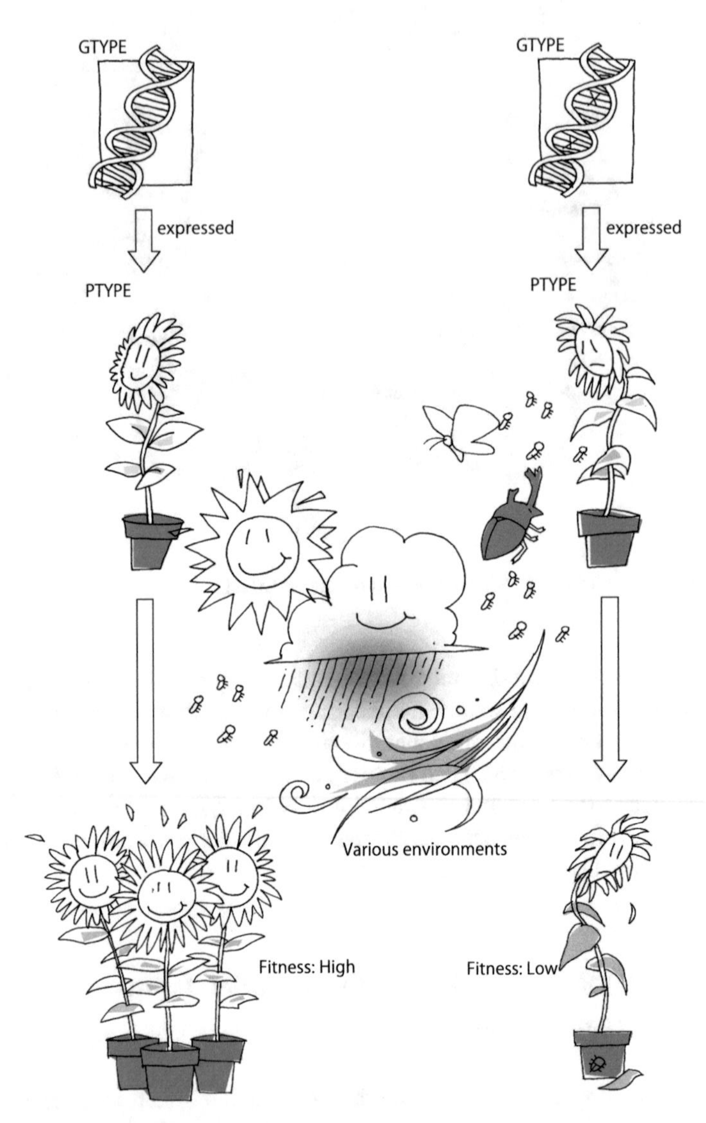

Fig. 1.5 GTYPE and PTYPE

and structures over a wide area, accompanied by development within a GTYPE environment. Fitness is determined by the PTYPE adapting to its environment, and selection relies on the fitness of the PTYPE (Fig. 1.5). For a time, the higher the fitness score taken, the better. Therefore, for individuals with a fitness of 1.0 and 0.3, the former can adapt better to their environment, and it is easier for them to survive (however, in other areas of this book, there are cases when it is better to have a smaller score).

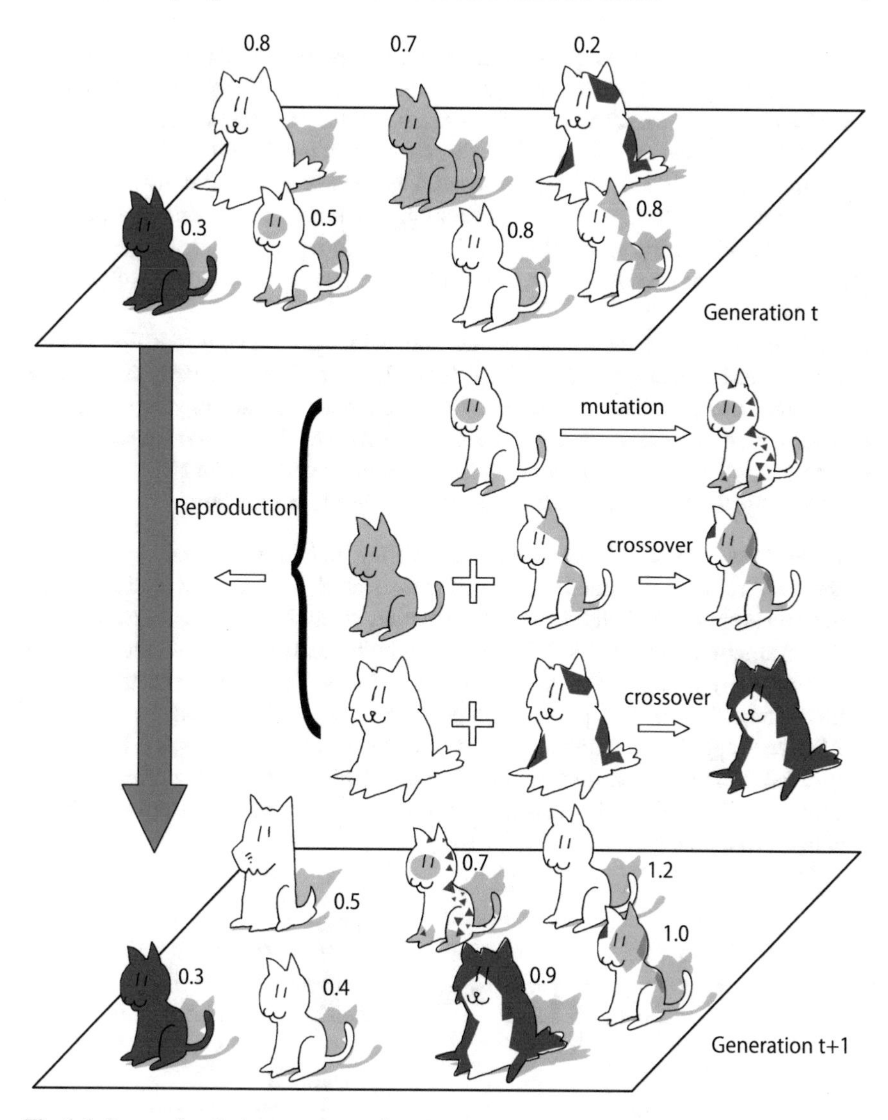

Fig. 1.6 Image of evolutionary computation

We shall explain the basic framework of the evolutionary computation, based on the above description (Fig. 1.6). Here, we configure a set containing several dogs. We shall call this generation t. This dog has a genetic code for each GTYPE, and its fitness is determined according to the generated PTYPE. In the diagram, the fitness of each dog is shown as the value near the dog (remember that the larger the better). These dogs reproduce and create the descendants in the next generation $t + 1$. In terms of reproduction, the better (higher) the fitness, the more descendants they are able to create, and the worse (lower) the fitness, the easier it is for them to become extinct (in biological terminology, this refers to choice and selection). In the diagram, the elements undergoing slight change in the phenotypes due to reproduction are drawn schematically. As a result of this, the fitness of each individual in the following generation $t + 1$ is expected to be better than that of the previous generation. Furthermore, the fitness as seen in the set as a whole also increases. In the same way, the dogs in the generation $t + 1$ become parents and produce the descendants in the generation $t + 2$. As this is repeated and the generations progress, the set as a whole improves and this is the basic mechanism of evolutionary computation.

In the case of reproduction, the operator shown in Fig. 1.7 is applied to the GTYPE and produces the next generation of GTYPEs. To simplify things, here, the GTYPE is expressed as a one-dimensional matrix. Each operator is an analogy for the genetic recombination and mutation, etc., in the organism. The application frequency and the application area of these operators are randomly determined in general.

Normally, the following kinds of methods are used for selection.

- **Roulette selection**: This is a method of selecting individuals in a ratio proportionate to their fitness. A roulette is created with an area proportionate to fitness. This roulette is spun, and individuals in the location where it lands are selected.
- **Tournament selection**: Only the number of individuals from within the set (tournament size) are chosen at random and individuals from these with the highest fitness are selected. This process is repeated for the number of sets.
- **Elite strategy**: Several individuals with the highest fitness are left as is to the next generation. This can prevent individuals with the highest fitness not being selected coincidentally and left to perish. This strategy is used in combination with the above two methods.

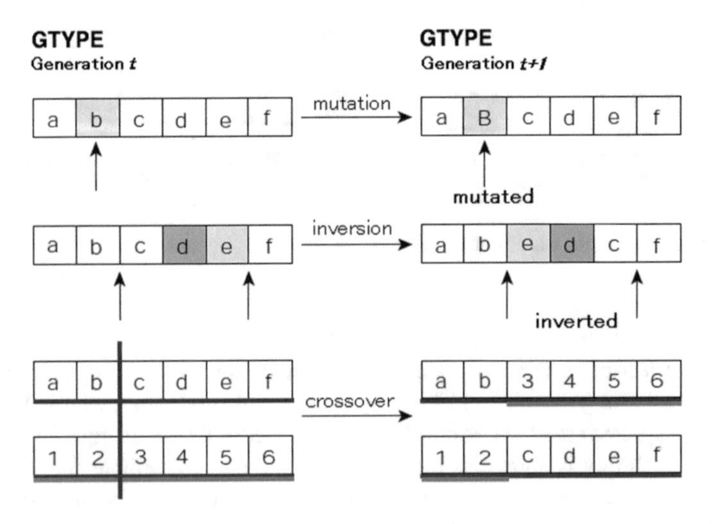

Fig. 1.7 Genetic operators

With the elite strategy, the results will not get worse in the next generation as long as the environment does not change. For this reason, it is frequently applied in engineering applications. However, note that the flip side of this is that diversity is lost. Related to this, an interesting experiment on chickens was conducted by the poultry researcher, William Muir [17]. In the modern poultry industry, approximately 10 or so hens are stuffed into a cage and raised. Let us try to improve the production volume of eggs through selective cultivation. Here, we shall introduce the following two methods (Fig. 1.8).

1. Choose the hens producing the most eggs from within the cage, and use them to breed the next generation of hens.
2. Select all hens within the cage with the highest volume of eggs produced, and use them to breed the next generation of hens.

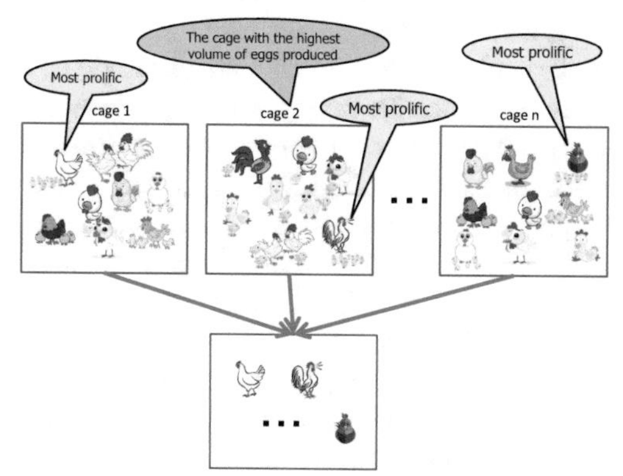

(a) Choose the hens producing the most eggs from within the cage.

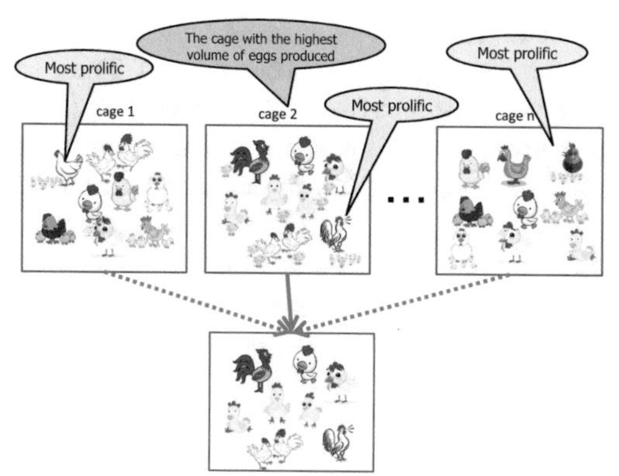

(b) Select all hens within the cage with the highest volume of eggs produced.

Fig. 1.8 How to improve the production volume of eggs?

Which of these two methods would be most effective? It is the individual hens that lay the eggs, so it would be expected that the first method that accumulates the best hens would be more effective than the best group, i.e., the second method. Note that the first method is regarded as the elite strategy. Muir experimented using both of these methods, and the results were very surprising. The result of producing the next generation by the first method meant that, after six generations, only three of the hens within the cage were left, and the other hens were slaughtered. The surviving three hens simply fought, had virtually no feathers, and were in no fit state to lay eggs. On the other hand, the hens selected by the second method were all well developed with plenty of feathers after six generations. Furthermore, the production quantity actually increased. Why was this kind of result produced?

This is because the hens laying the most eggs, by suppressing the egg production of other hens in the cage, were able to produce a lot. In the first method, the worst hen in each cage was selected, causing a tragic result. In this way, the elite strategy is a double-edged sword, and unless it is used correctly, it suppresses the diversity of the group and brings about an even worse result.

To summarize, in the evolutionary computation, generational change is as shown in Fig. 1.9. In the figure, G is the elite rate (rate of upper-level individuals that were copied and left results). We can refer to the reproduction rate as $1 - G$.

We shall now introduce some interesting research on evolutionary computation using actual living organisms [3, 4]. Bond et. al. used blue jays[7] and experimented in regard to the evolution in patterns of moths, which are their food. To do this, they made actual blue jays attack the digital images of moths projected on computer screens (see Fig. 1.10). On each experiment date, they removed the moths that had been discovered (attacked) by the blue jays. Then, the number of moths surviving in each set was restored, thus maintaining it at a regular ratio the following morning (generational change proportionate to survival rate). On repeating the experiment for 30 days, that is to say 30 generations, a hidden type of moth stabilized at a ratio of 75%.

Next, they introduced a mutation or crossover during the generational change, and the genetics of brightness and pattern were evolved. The result of this was that the blue jay frequently failed to discover the anomalous hidden type of moth due to mutation. For this reason, the variants increased in frequency. As the generations passed, the moths become more difficult to detect and this phenotype (patterned) demonstrated major disparities (see Fig. 1.11; the patterns of the moths are shown on the left, while the right shows the state presented to the blue jays). This is similar to coevolution (arms races, see p. 4) occurring in the natural world. In actual fact, prey animals improve concealment and mimicry patterns to make themselves harder for predators to find and are observed to increase in multiple forms.

[7]Birds in the sparrow family, with brilliant blue tail feathers. They mainly feed on insects, seeds, and fruit.

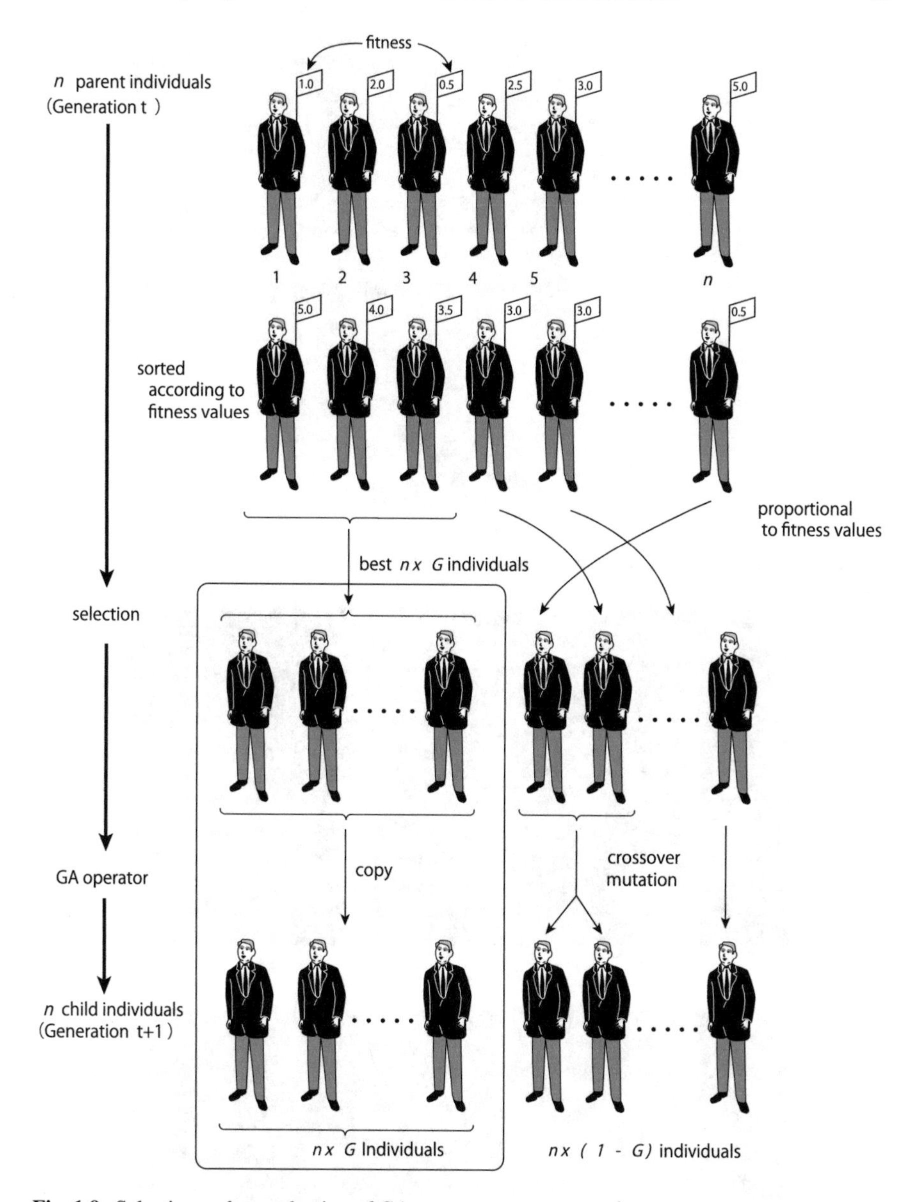

Fig. 1.9 Selection and reproduction of GA

There are many riddles remaining biologically in terms of the evolution of mimicry. Research into solving these mysteries is flourishing with the use of computer simulation using evolutional computation.

Evolutional computation is used in a variety of areas of our everyday lives. For example, the front carriage model of the Japanese N700 series bullet train plays a

Fig. 1.10 Images of moths selected by actual blue jays

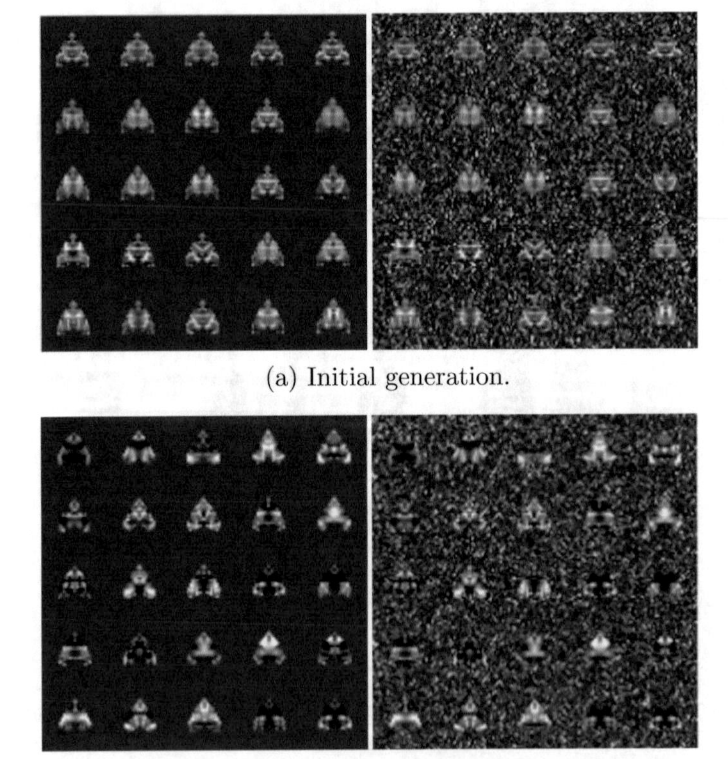

Fig. 1.11 Evolution of the patterns of moths (Reproduced from [4])

(a) N700 series bullet train.

(b) MRJ (Mitsubishi regional jet).

(c) Collaborative transportation by humanoid robots

Fig. 1.12 EC applications

major role in creating original forms (Fig. 1.12a). The N700 has the performance to take curves at 270 km, speeds 20 km faster than the previous model. However, in the traditional form of the front carriage, speeding up meant that the microbarometic waves in the tunnel increased, which are a cause of noise. To solve this difficulty, the original form known as "Aero Double Wing" has been derived from approximately 5,000 simulations using evolutionary computation. Furthermore, in the wing design of the MRJ (Mitsubishi regional jet), which is the first domestic jet in Japan, a method known as multi-objective evolutionary computation was used (Fig. 1.12b). Using this method, the two objectives of improving the fuel efficiency of passenger jet devices and reduction in noise external to the engine were optimized simultaneously, and they succeeded in improving performance compared to competing models.

In fields other than engineering, such as the financial field, the use of evolutionary computation methods is spreading. Investment funds are using this as a practical technology for portfolio construction and market prediction (see [7] for details). Furthermore, it has practical application in such fields as scheduling design to optimize the work shifts of nurses and allocating crews for aircraft.

Another field that is using evolutionary computation is the field of evolutionary robotics (see also Sect. 5.3.1). For example, Fig. 1.12c is an example of cooperative work (collaborative transportation) of evolutionary humanoid robots. Here, a learning model is used that applies coevolution to evolutionary computation. Furthermore, module robots, which modify themselves in accordance with geographical features, environment, and work content, by combining blocks, are gaining attention. This technology is even being used by NASA (National Aeronautics and Space Administration) for researching the form of robots optimized for surveying amidst the limited environment of Mars. The form of organisms we know about may be only those species that are remaining on earth. These may be types that match the earth environment, and it is not known if these are optimal. Through evolutionary computation, if we can reproduce the process of evolution on a computer, new forms may emerge that we do not yet know about. The result of this may be the evolution of robots compatible with Mars and unknown planets (see [9]).

1.4 Genetic Programming and Its Genome Representation

1.4.1 Tree-Based Representation of Genetic Programming

The aim of Genetic Programming (GP) is to extend genetic forms from Genetic Algorithm (GA) to the expression of trees and graphs and to apply them to the synthesis of programs and the formation of hypotheses or concepts. Researchers are using GP to attempt to improve their software for the design of control systems and structures for robots.

The procedures of GA are extended in GP in order to handle graph structures (in particular, tree structures). Tree structures are generally well described by S-expressions in LISP. Thus, it is quite common to handle LISP programs as "genes" in GP. As long as the user understands that the program is expressed in a tree format, then he or she should have little trouble reading a LISP program (the user should recall the principles of flowcharts). The explanations below have been presented so as to be quickly understood by a reader who does not know LISP.

A tree is a graph with a structure as follows, incorporating no cycles:

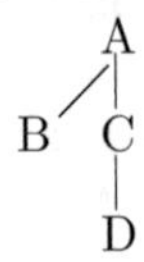

More precisely, a tree is an acyclical connected graph, with one node defined as the root of the tree. A tree structure can be expressed as an expression with parentheses. The above tree would be written as follows:

```
(A (B)
   (C (D)))
```

In addition, the above can be simplified to the following expression:

```
(A B
  (C D))
```

This notation is called an "S-expression" in LISP. Hereinafter, a tree structure will be identified with its corresponding S-expression. The following terms will be used for the tree structure:

- Node: Symbolized with A, B, C, D, etc,
- Root: A,
- Terminal node: B, D (also called a "terminal symbol" or "leaf node"),
- Non-terminal node: A, C (also called a "non-terminal symbol" and an "argument of the S-expression"),

- Child: From the viewpoint of A, nodes B and C are children (also, "arguments of function A"),
- Parent: The parent of C is A.

Other common phrases will also be used as convenient, including "number of children", "number of arguments", "grandchild", "descendant", and "ancestor." These are not explained here, as their meanings should be clear from the context.

The following genetic operators acting on the tree structure will be incorporated:

1. **Gmutation** Alteration of the node label,
2. **Ginversion** Reordering of siblings,
3. **Gcrossover** Exchange of a subtree.

These are natural extensions of existing genetic operators and act on sequences of bits. These operators are shown below in examples where they have been applied in LISP expression trees (S-expressions) (Fig. 1.13). The underlined portion of the statement is the expression that is acted upon:

Gmutation Parent :(+ x $\underline{y}$)

⇓

Child :(+ x $\underline{z}$)

Ginversion Parent :(progn $\underline{\text{(incf } x\text{) (setq } x \text{ 2)}}$ (print x))

⇓

Child :(progn $\underline{\text{(setq } x \text{ 2) (incf } x\text{)}}$ (print x))

Gcrossover Parent$_1$:(progn (incf x) $\underline{\text{(setq } x \text{ 2)}}$ (setq y x))

Parent$_2$:(progn (decf x) (setq x (* $\underline{\text{(sqrt } x\text{)}}$ x)) (print x))

⇓

Child$_1$:(progn (incf x) $\underline{\text{(sqrt } x\text{)}}$ (setq y x))

Child$_2$:(progn (decf x) (setq x (* $\underline{\text{(setq } x \text{ 2)}}$ x)) (print x))

Table 1.1 provides a summary of how the program was changed as a result of these operators. "progn" is a function acting on the arguments in the order of their presentation and returns the value of the final argument. The function "setq" sets the value of the first argument to the evaluated value of the second argument. It is apparent on examining this table that mutation has caused a slight change to the action of the program, and that crossover has caused replacement of the actions in parts of the programs of all of the parents. The actions of the genetic operators have produced programs that are individual children but that have inherited the characteristics of the parent programs.

1.4.2 Linear Genetic Programming

In a linear version of GP, variables and constants are considered to be a function which returns a certain value so that we can treat both terminal and functional nodes

Gmutation

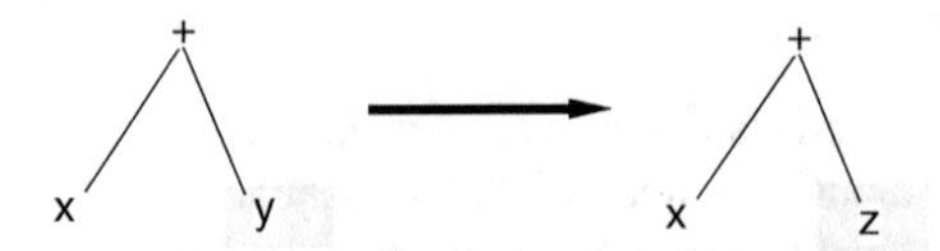

Ginversion

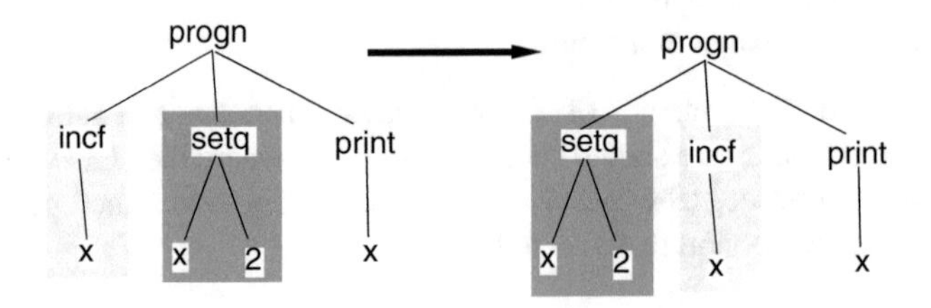

Gcrossover

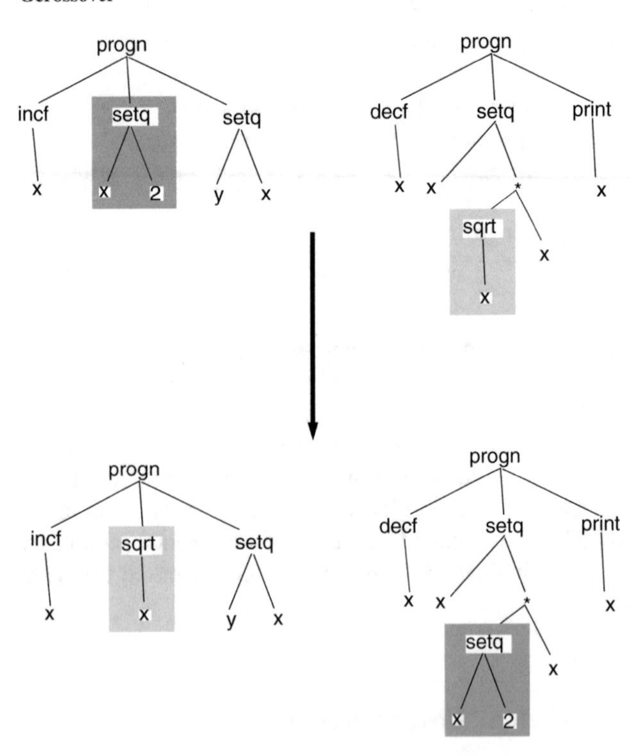

Fig. 1.13 Genetic operators in GP

Table 1.1 Program changes due to GP operators

Operator	Program before operation	Program after operation
Mutation	Add x and y	Add x and z
Inversion	1. Add 1 to x	1. Set $x = 2$
	2. Set $x = 2$	2. Add 1 to x
	3. Print $x(=2)$ and return 2	3. Print $x(=3)$ and return 3
Crossover	Parent_1	Child_1
	1. Add 1 to x	1. Add 1 to x
	2. Set $x = 2$	2. Take square root of x
	3. Set $y = x(= 2)$ and return 2	3. Set $y = x$ and return the value
	Parent_2	Child_2
	1. Subtract 1 from x	1. Subtract 1 from x
	2. Set $x = \sqrt{x} \times x$	2. Set $x = 2$ and its value (=2) is multiplied by $x(=2)$. The result value (=4) is set to x again
	3. Print x and return the value	3. Print $x(=4)$ and return 4

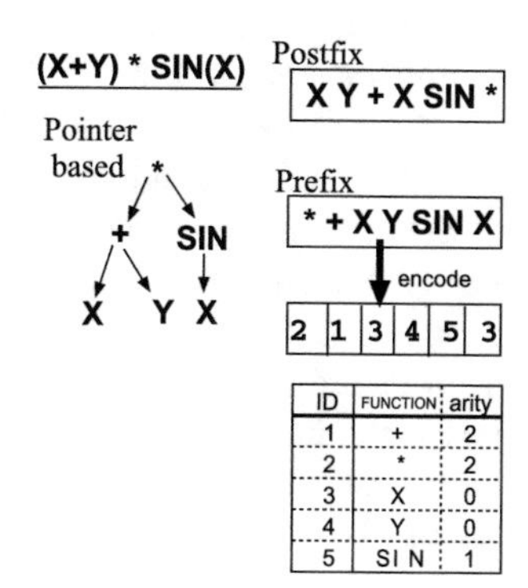

Fig. 1.14 Linear representation of a GP individual

in the same way. We assign an ID number to the function in order to distinguish one another. GP individuals are represented as an array, in which the ID number of each tree node is put in the prefix order (see Fig. 1.14 for example). When a genome is being evaluated, the ID number of a node is used as the index of the function table. The table retains the information on the function such as the number of arguments and pointers to the definition of functions.

When we implement genetic operators such as a crossover, the most important is that it should not generate structurally wrong children, i.e., lethal genes. This is easily secured in the pointer-based implementation, which explicitly represents a tree structure with pointers, because the crossover in such system is executed by the exchange of pointers that refer to subtrees to be exchanged (see Fig. 1.13). But it is not so simple in GP systems with the implicit linear genome representation. In order to apply a crossover to a linear genome, we have to determine the exact range of a subtree on the genome at first.

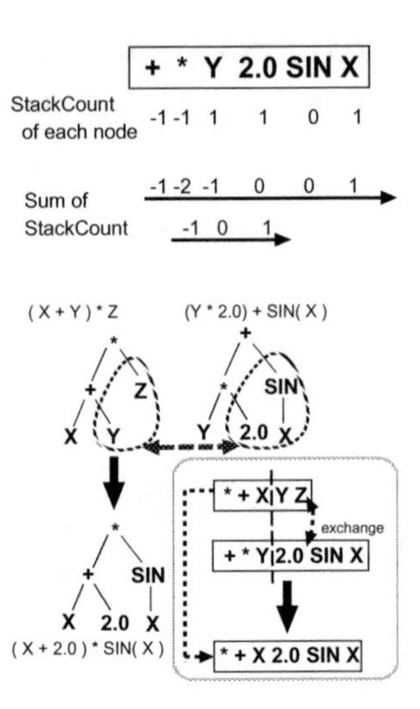

Fig. 1.15 Introduction of StackCount

Fig. 1.16 One-point crossover in linear GP

Keith and Martin [8] introduced the idea of "StackCount" for the sake of such a structure analysis (Fig. 1.15). Linear genomes can be executed by pushing on and popping off the stack. StackCount for a node equals the number of arguments it pushes on the stack minus the number of arguments it pops off the stack. The sum of StackCount is always less than one except at the end of a genome as we scan from left to right on a structurally correct tree and a subtree represented as the linear genome in the prefix order. The total sum of StackCount must be one at the gene end. With this rule, we can determine the end of a subtree. The idea of the StackCount makes the initialization of the GP population simple and fast. In order to initialize a gene, we randomly select a node and put it to an array in sequence till the sum of the StackCount becomes one. This implementation can omit the recursive pointer references required for the pointer-based tree representation.

The StackCount rule is similarly used for the implementation of the GA-like one-point crossover. If the ongoing StackCount values are equal just before any two loci on two parents, the loci can be crossover points. This restriction prevents the generation of lethal genes when applying a one-point crossover. In terms of the topology of GP individuals, the application of a one-point crossover is to divide a tree structure into two part obliquely and to exchange it between the parent trees (Fig. 1.16). If we use the conventional subtree exchange crossover, the same effect is possible only when it is applied repeatedly.

Therefore, the one-point crossover can have a greater influence on the semantic alternation of genes, preventing the premature convergence to a local optimum. The one-point crossover works on a tree structure asymmetrically. Thus, the efficiency of the operator is supposed to depend on the property of the problem and the function/terminal set. The effectiveness of this operator is further discussed in [15].

Prefix genes can be executed by evaluating each node recursively from the top of an array. A global variable indicates which node of the array is being evaluated. It is

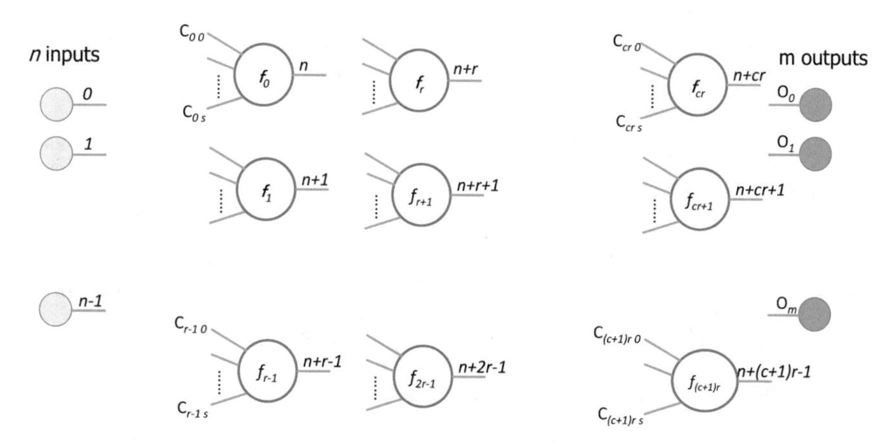

Fig. 1.17 Example of a CGP network

incrementally updated after the function for the indicated node is called. Genes are executed in this manner in our system. In the prefix ordering, functions are evaluated before their arguments and the tree structure of genes can be analyzed by scanning the StackCount values of the array to the end. As a result of this, we can implement the flow control mechanisms, such as conditional statements and iterations, more easily.

1.5 Cartesian Genetic Programming (CGP)

CGP [10] is a Genetic Programming (GP) technique proposed by Miller et al. CGP represents a tree structure with a feed-forward-type network. It is a method by which all nodes are described genotypically beforehand to optimize connection relations. This is supposed to enable handling the problem of bloat, in which the tree structure becomes too large as a consequence of the number of GP genetic operations. Furthermore, by reusing a partial tree, the tree structure can be represented compactly.

The CGP network comprises three varieties of node: input nodes, intermediate nodes, and output nodes. Figure 1.17 shows a CGP configuration with n inputs, m outputs, and $r \times c$ intermediate layers. Here, connecting nodes in the same column is not permitted. CGP networks are also restricted to being feed-forward networks.

CGP uses a one-dimensional numeric string for the genotypes. These describe the functional type and connection method of the intermediate nodes, and the connection method of the output nodes. Normally, all functions have the largest argument as the input and ignore unused connections. For example, consider the following genotypes for the CGP configuration shown in Fig. 1.18.

```
0 0 1   1 0 0   1 3 1   2 0 1   0 4 4   2 5 4     2 5 7 3
```

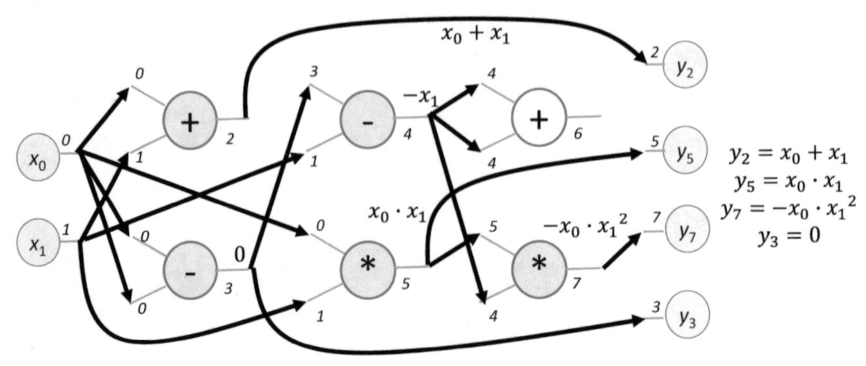

Fig. 1.18 Example of a genotype and a phenotype

The function symbol numbers 0, 1, 2, and 3 (underlined above) correspond to addition, subtraction, multiplication, and division, respectively. The network corresponding to the genotype at this time is as shown in Fig. 1.18. For example, the inputs of the first node 0 are input 0 and input 1, and the addition computation is functional. Note that the output of the fifth node is not used anywhere (0 4 4), making it an intron (i.e., a noncoding region).

1.6 Interactive Evolutionary Computation (IEC)

Let us consider how to use evolutionary computation for problems such as "designing a table that matches the atmosphere of the room" or "formulating a mobile phone ringtone that does not annoy other people." The problem at this time is how to calculate the evaluation (fitness) of each individual. Having a computer automatically evaluate whether a certain table matches the atmosphere of the room is certainly not easy. In particular, it is extremely difficult to model subjective judgments based on human preferences and feeling and implement these on a computer. As feelings differ according to the individual, there are no functions that can be optimized to produce music or design that is liked by large quantities of people. Therefore, we shall take an evolutionary approach that incorporates an optimized system for the evaluation system of humans as an evaluation function. That is to say, a method in which humans directly evaluate each individual. In this way, the evolutionary computation for optimizing based on human subjective evaluation is called the interactive evolutionary computation (IEC) method.

In IEC, human evaluation substitutes for the fitness derivation in EC. As shown in Fig. 1.19, with IEC, the user directly evaluates each individual. That is to say, the survival level (fitness) of the next generation is determined according to the "preferences" of the user. In this way, rather than modeling the preferences and feelings of the individual, an evaluation system is integrated within the system based

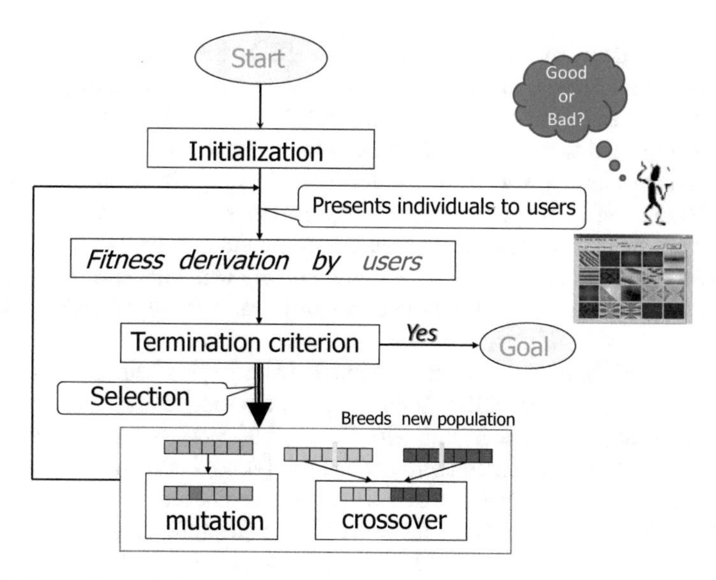

Fig. 1.19 Interactive evolutionary computation

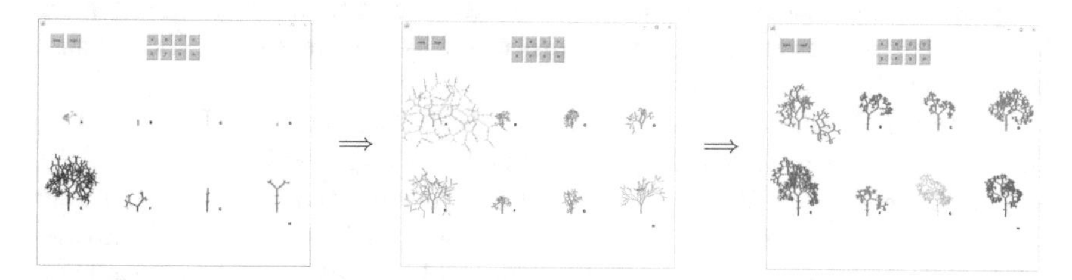

Fig. 1.20 Trees evolved by IEC

on the subjective perspectives of the user, as a black box. Whereas the original evolutionary computation modeled the results of survival competition for life, IEC is a method of obtaining hints for improving the quality of the agriculture and livestock carried out by humans. This process of evaluation is the same as that of the females in sexual selection. By incorporating this into the evaluation loop in Fig. 1.19, it is expected that the user's thinking will be stimulated and changed. As a result of this, we can expect the evolution of the evaluator (user) as well.

We attempted to create a plant CG synthesis demo using IEC (Fig. 1.20). With this demo, we generate a tree structure based on the L-system.[8] The parameters for generating the tree and the designation of colors are a GTYPE in IEC. Eight trees are presented to the user. From there, where several trees are chosen and clicked, these become parents that generate children. When this system is used, it is possible to experience the enjoyment of "raising" your favorite plant.

[8]See Sect. 3.1 for the details of L-system.

IEC is applied to various fields. For example, IEC-based automatic correction for digital hearing aids [14] is an application for "human-friendly technology" that aims to integrate human senses and subjectivity into the system. With the IEC format, a method is proposed to regulate signal processing parameters based on how well the users themselves hear. It was reported that approximately 70% of users replied that this was significantly better compared to the previous format (a method in which specialist physician and technicians regulate the parameters based on their hearing tests). The expectations for IEC are the highest for such engineering applications dealing with the sensations and preferences of users that are difficult to measure quantitatively.

A famous IEC application related to music is an improvised jazz performance system, GenJam [2] using MIDI. With GenJam, the individuals in the population combine and play melodies in line with the rhythm sections and chords input in advance. Evaluators listening to it pressed the "g" (good) button or "b" (bad) button at that instant and evaluated each item. As generations proceed based on the provided evaluations, individuals with better (at least closer to the preferences of the evaluator) tremolos and phrases are created. This equates to melody ideas being accumulated by jazz musicians through repeated trial and error. The veteran jazz performer Biles, the producer of GenJam, is performing jam sessions with GenJam, refined in this way. GenJam repeats the kind of performance that humans would not normally think of, so it frequently surprises us.

Another IEC-based composing system is CACIE (Computer-Aided Composition System by means of Interactive EC). Figure 1.21 is an interface for the CACIE system [1]. Each respective ball has a specific melody, and similar melodies are expressed using similar colors. The composer places the balls for the melodies they like inside and the balls for those that they do not like outside. While repeating the generational changes over several generations, the various balls inherit the traits of their parent generation, and different melodies are generated through mutation. The more generations it passes, the more refined the song becomes. One excellent point of this system is that people who do not know about music can create interesting songs.[9] In other words, a song can be completed just by listening to melodies and deciding whether you like or dislike them. Even if they are unable to write score, as long as they have a vague image of what they wish to do they can create a song. This is just drawing out your own preferences from many different solutions. In actual fact, when this system was operated by piano beginners, music that was easy to perform, such as Bayern, could be created.

On the other hand, when CACIE was used by people who were very knowledgeable about music, exquisite harmonies could be drawn out, and the content was entered into several contests.[10] It was learned that, depending on the person using it,

[9]The CACIE demonstration was broadcasted on TV channels of Japan as well as on the satellite broadcasting.

[10]The contents composed with CACIE have been awarded several times in computer music conferences, such as ICMC (International Computer Music Conference), and played by professional musicians.

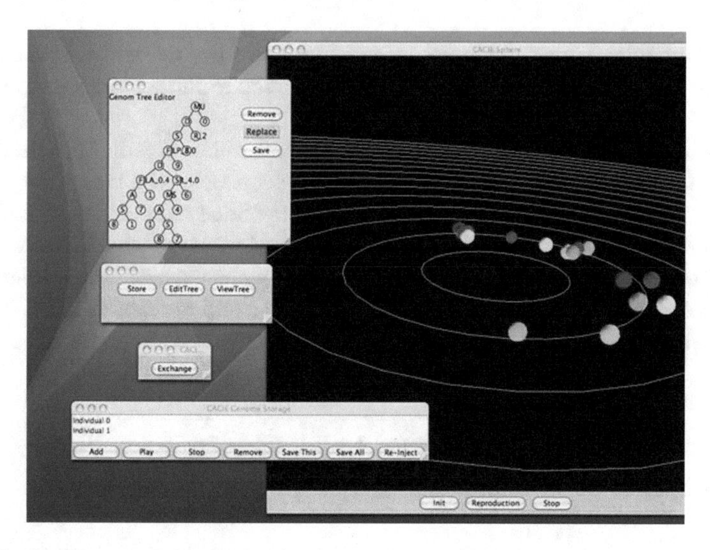

Fig. 1.21 CACIE composing system

the songs completed were surprisingly different. This is really an important point in interactive evolutionary computation. The fitness requested by the system is determined according to the person carrying out the input work. The result of this is that humans become integrated with the system, and production is performed with one's own thinking stimulated by mutation.

1.7 Why Evolutionary Computation?

Evolutionary computation has different advantages than other AI methods and classical optimization methods. In particular, it is a major strength that it can efficiently search for combinations of a large number of candidates. As has been seen until now, evolutionary computation has a high level of versatility, and as long as

- Converting from GTYPE to PTYPE
- Fitness functions
- Genetic operators

can be appropriately determined, it can be applied to problems without requiring excessive effort. This can be improved as necessary after application and realistically improve performance. For this reason, it may be effective in evolving solutions of which humans cannot conceive.

Contrary to the neural network (or simulated annealing method), as it can search using a combination of structures, unexpected solutions can be achieved. This is one of the highlights of using evolutionary computation. This may be effective if it can evolve solutions and structures that cannot be thought of by humans. In the case of

IEC, in particular, concepts presenting non-forecasted and unexpected concepts can frequently be stimulated, and this has been observed to further deepen the creative process.

The field of EA is characterized by strong collaboration with other fields, and EA contains research areas with projects that attract people with versatility, competitiveness, and harmony. As Melanie Mitchell has pointed out [11], "Just as GAs are "general-purpose" search methods, GA researchers have to be generalists, willing to step out of their own discipline and learn something about a new one in order to pursue a promising application or model." One of the exciting features of EC research is the ability to interact with various fields, and as a result it allows for collaborative research to be gradually initiated with scientists from different fields.

References

1. Ando, D., Iba, H.: Interactive composition aid system by means of tree representation of musical phrase. In: Proceedings of IEEE Congress on Evolutionary Computation, pp. 4258–4265 (2007)
2. Biles, J.A.: Life with GenJam: interacting with a musical IGA. In: Proceedings of the 1999 IEEE International Conference on Systems, Man, and Cybernetics, pp. 652–656 (1999)
3. Bond, A.B., Kamil, A.C.: Apostatic selection by blue jays produces balanced polymorphism in virtual prey. Nature **395**, 594–596 (1998)
4. Bond, A.B., Kamil, A.C.: Visual predators select for crypticity and polymorphism in virtual prey. Nature **415**, 609–613 (2002)
5. Dawkins, R.: Climbing Mount Improbable. W. W. Norton & Company, New York (1997)
6. Grant, P.R., Grant, B.R.: How and Why Species Multiply: The Radiation of Darwin's Finches. Princeton University Press, Princeton (2011)
7. Iba, H., Aranha, C.C.: Practical Applications of Evolutionary Computation to Financial Engineering: Robust Techniques for Forecasting, Trading and Hedging. Springer, Berlin (2012)
8. Keith, M.J., Martin, M.C.: Genetic programming in C++: implementation issues. In: Kinnear Jr., K.E. (ed.) Advances in Genetic Programming, pp. 285–310. MIT Press, Cambridge (1994)
9. Lipson, H., Pollack, J.B.: Automatic design and manufacture of robotic lifeforms. Nature **406**, 974–978 (2000)
10. Miller, J.F. (ed.): Cartesian Genetic Programming. Springer, Berlin (2011)
11. Mitchell, M.: An Introduction to Genetic Algorithms. MIT Press, Cambridge (1998)
12. Nilsson, D.-E., Pelger, S.: A pessimistic estimate of the time required for an eye to evolve. Proc.: Biol. Sci. **256**(1345), 53–58 (1994)
13. Parker, A.: In the Blink of an Eye: How Vision Sparked the Big Bang of Evolution. Basic Books, New York (2004)
14. Takagi, H., Ohsaki, M.: IEC-based hearing aids fitting. In: Proceedings of the IEEE International Conference on Systems, Man, and Cybernetics (IEEE SMC99), pp. 657–662 (1999)
15. Tokui, N., Iba, H.: Empirical and statistical analysis of genetic programming with linear genome. In: Proceedings of the Systems, Man, and Cybernetics Conference (SMC'99), pp. 610–615 (1999)
16. Weiner, J.: The Beak of the Finch: A Story of Evolution in Our Time. Vintage, New York (1995)
17. Wilson, D.S.: Evolution for Everyone: How Darwin's Theory Can Change the Way We Think About Our Lives. Delacorte Press, New York (2007)

Chapter 2
Meta-heuristics, Machine Learning, and Deep Learning Methods

I remember the first time I met Edsger Dijkstra. ... He asked me what I was working on. Perhaps just to provoke a memorable exchange I said, "AI." To that he immediately responded, "Why don't you work on I?"

(Leslie Valiant, Probably Approximately Correct: Nature's Algorithms for Learning and Prospering in a Complex, Basic Books 2014)

Abstract This chapter introduces several meta-heuristics and learning methods, which will be employed in later chapters. These methods will be employed to extend evolutionary computation frameworks in later chapters. Readers familiar with these methods may skip this chapter.

Keywords Particle swarm optimization (PSO) · Differential evolution (DE) k-means algorithm · Support vector machine (SVM) · Relevance vector machine (RVM) · k-nearest neighbor classifier (k-NN) · Transfer learning · Bagging Boosting · Gröbner bases · Affinity propagation · Convolutional neural networks (CNN) · Generative adversary networks (GAN) · Bayesian networks · Loopy belief propagation

2.1 Meta-heuristics Methodologies

2.1.1 PSO: Particle Swarm Optimization

"Particle swarm optimization" (PSO) is an algorithm from the field of swarm intelligence. It was first described by Eberhart and Kennedy as an alternative to GA in 1995 [36, 37]. The algorithm for PSO was conceived on the basis of observations of certain social behavior in lower-class animals or insects. In contrast to the concept of

H. Iba, *Evolutionary Approach to Machine Learning and Deep Neural Networks*, https://doi.org/10.1007/978-981-13-0200-8_2

modifying genetic codes using genetic operators as used in GA, in PSO the moving individuals (called "particles") are considered where the next movement of an individual is determined by the motion of the individual itself and that of the surrounding individuals. It has been established that PSO has capabilities equal to those of GA for function optimization problems. There have been several comparative studies on PSO and standard GA (see [2, 19, 27, 38]).

The classic PSO was intended to be applied to optimization problems. It simulates the motion of a large number of individuals (or "particles") moving in a multidimensional space [36]. Each individual stores its own location vector ($\vec{x}_i$), velocity vector ($\vec{v}_i$), and the position at which the individual obtained the highest fitness value ($\vec{p}_i$). All individuals also share information regarding the position with the highest fitness value for the group ($\vec{p}_g$).

As generations progress, the velocity of each individual is updated using the best overall location obtained up to the current time for the entire group and the best locations obtained up to the current time for that individual. This update is performed using the following formula:

$$\vec{v}_i = \chi \cdot (\omega \cdot \vec{v}_i + \phi_1 \cdot (\vec{p}_i - \vec{x}_i) + \phi_2 \cdot (\vec{p}_g - \vec{x}_i)) \tag{2.1}$$

The coefficients employed here are the convergence coefficient χ (a random value between 0.9 and 1.0) and the attenuation coefficient ω, while ϕ_1 and ϕ_2 are random values unique to each individual and the dimension, with a maximum value of 2. When the calculated velocity exceeds some limit, it is replaced by a maximum velocity V_{max}. This procedure allows us to hold the individuals within the search region during the search.

The locations of each of the individuals are updated at each generation by the following formula:

$$\vec{x}_i = \vec{x}_i + \vec{v}_i \tag{2.2}$$

The overall flow of the PSO is as shown in Fig. 2.1. Let us now consider the specific movements of each individual (see Fig. 2.2). A flock consisting of a number of birds is assumed to be in flight. We focus on one of the individuals (Step 1). In the figure, the ○ symbols and linking line segments indicate the positions and paths of the bird. The nearby ◎ symbol (on its path) indicates the position with the highest fitness value on the individual's path (Step 2). The distant ◎ symbol (on the other bird's path) marks the position with the highest fitness value for the flock (Step 2). One would expect that the next state will be reached in the direction shown by the arrows in Step 3. Vector ① shows the direction followed in the previous steps; vector ② is directed toward the position with the highest fitness for the flock; and vector ③ points to the location where the individual obtained its highest fitness value so far. Thus, all these vectors, ①, ②, and ③, in Step 3 are summed to obtain the actual direction of movement in the subsequent step (see Step 4).

The efficiency of this type of PSO search is certainly high because focused searching is available near-optimal solutions in a relatively simple search space. However, the canonical PSO algorithm often gets trapped in local optimum in multimodal

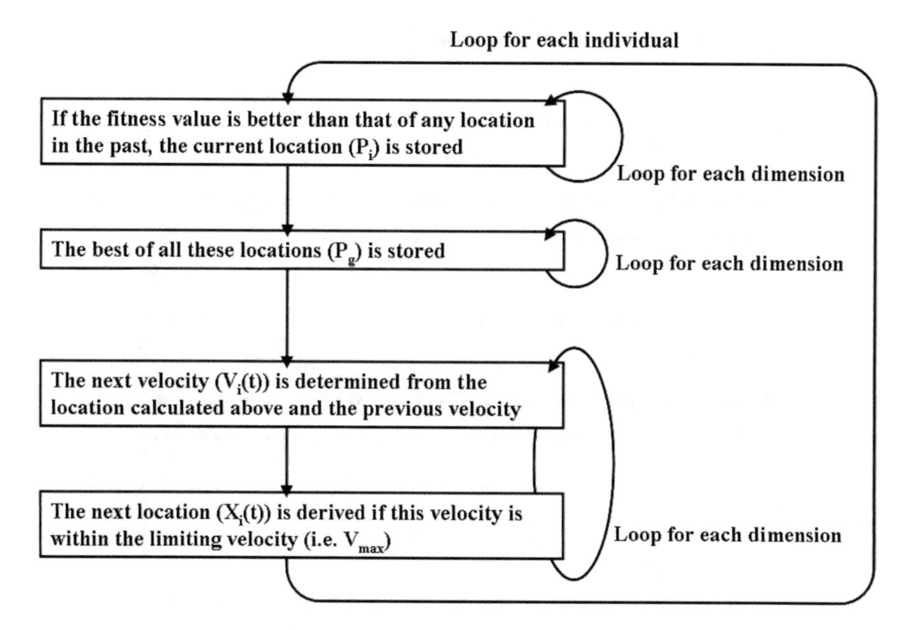

Fig. 2.1 Flow chart of the PSO algorithm

problems. Because of that, some sort of adaptation is necessary in order to apply PSO to problems with multiple sharp peaks.

To overcome the above limitation, a GA-like mutation can be integrated with PSO [27]. This hybrid PSO does not follow the process by which every individual of the simple PSO moves to another position inside the search area with a predetermined probability without being affected by other individuals, but leaves a certain ambiguity in the transition to the next generation due to Gaussian mutation. The individuals are selected at a predetermined probability and their positions are determined at the probability under the Gaussian distribution. Wide-ranging searches are possible at the initial search stage and search efficiency is improved at the middle and final stages by gradually reducing the appearance ratio of Gaussian mutation at the initial stage. Figure 2.3 shows the PSO search process with Gaussian mutation. In the figure, V_{lbest} represents the velocity based on the local best, i.e., $\vec{p}_i - \vec{x}_i$ in Eq. (2.1), whereas V_{gbest} represents the velocity based on the global best, i.e., $\vec{p}_g - \vec{x}_i$.

PSO is a stochastic search method, as are GA and GP, and its method of adjustment of $\vec{p}_i$ and $\vec{p}_g$ resembles crossover in GA. It also employs the concept of fitness, as in evolutionary computation. Thus, the PSO algorithm is strongly related to evolutionary computation (EC) methods. In conceptual terms, one could place PSO somewhere between GA and EP.

However, PSO has certain characteristics that other EC techniques do not have. GA operators directly operate on the search points in a multidimensional search space, while PSO operates on the motion vectors of particles which in turn update the search points (i.e., particle positions). In other words, genetic operators are position specific

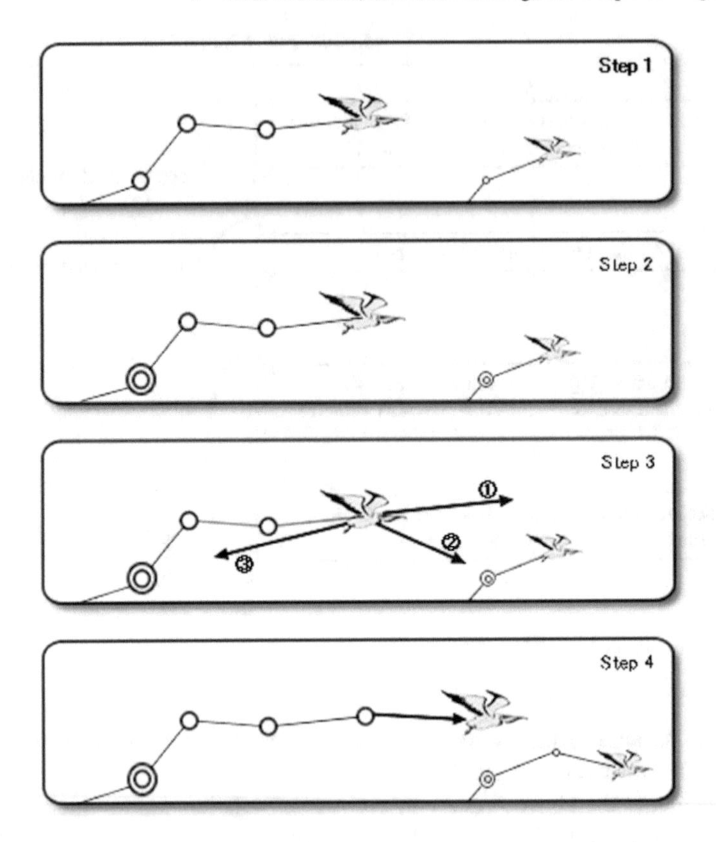

Fig. 2.2 In which way do birds fly?

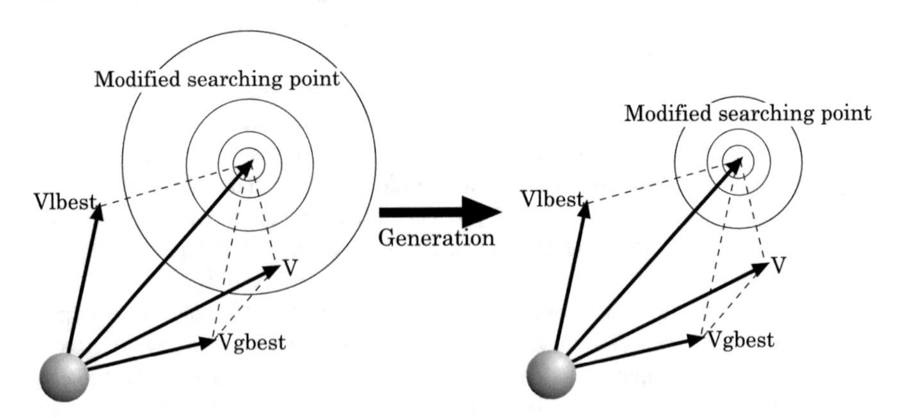

Fig. 2.3 Concept of searching process by PSO with Gaussian mutation

and the PSO operators are direction specific. One of the reasons PSO has gathered so much attention was the tendency of its individuals to proceed directly toward the target.

In the chapter "The Optimal Allocation of Trials" in his book [29], Holland ascribes the success of EC on the balance of "exploitation," through search of known regions, with "exploration," through search, at finite risks, of unknown regions. PSO is adept at managing such subtle balances. These stochastic factors enable PSO to make thorough searches of the relatively promising regions and, due to the momentum of speed, also allows effective searches of unknown regions.

2.1.2 DE: Differential Evolution

The subsection presents another EA called differential evolution (DE) for real parameter optimization. In the field of continuous optimization, the DE algorithm has drawn much attention in the last few years after it was proposed by Rainer M. Storn and Kenneth V. Price in 1995 [65]. Being a member of the EA family, it works with a population of solutions and stochastically searches through the search space. The beauty of the algorithm is its simple and compact structure which is very easy to understand and work with. In spite of its primitive architecture, it exhibits a powerful search capability which has brought it many real-world applications in diverse fields. Storn and Price proposed DE as a family of algorithms rather than as a single algorithm [65, 66].

Differential evolution was the outcome of the research by Kenneth Price and Rainer Storn for solving the Chebyshev polynomial fitting problem. Using common concepts of EAs, they converted a combinatorial algorithm (the genetic annealing algorithm developed by Kenneth Price) to a numerical optimizer. Therefore, DE works with the general framework of EA and uses many of EA concepts such as multipoint searching, use of recombination, and selection operators.

Like most of the EAs, DE starts to explore the search space by sampling at multiple, randomly chosen initial points [59, 64]. Thereafter, the algorithm guides the population toward the vicinity of the global optimum through repeated cycles of reproduction and selection. The generation alternation model used in "classic DE" for refining candidate solutions in successive generations is almost the same as the one used in GA or GP frameworks. Looking at the overall framework of DE, it is not easy to differentiate the algorithm from any ordinary GA because it iterates with the same common components: *initialization*, *evaluation*, *reproduction*, and *selection*. However, as we look inside these components in the following sections the distinct features of the algorithm will become apparent.

One important concern in implementing any EA is the initialization of the population. Actually, there are two issues here: "How to initialize each gene of the individual?" and "How many individuals to be used in the population?" Here we will discuss only the first issue. See [33, Sect. 3.4.7.3] for more details of the other issue, which is related to its population size, a critical parameter of DE etc.

EA algorithms usually initialize each gene of each chromosome (individual) using a uniform random generator within the search ranges. DE is not an exception in this regard. Suppose we are working in an N-dimensional problem. Then each individual of the DE population, P^G, would be an N-dimensional vector which can be initialized as follows:

$$x_{i,t}^{G=1} = Random_t(LB_t, UB_t), \tag{2.3}$$

where $x_{i,t}^{G=1}$ denotes the tth gene ($t = 1, 2, \ldots, N$) of ith individual ($i = 1, 2, \ldots, P$) in generation $G = 1$. LB_t and UB_t denote the lower and upper limit of the search ranges for gene-j, respectively, and $Random_t(a, b)$ denotes the uniform random number generator that returns a uniformly distributed random number from [a, b). The subscript in $Random_t$ is used to clarify that a separate random number is drawn for each gene in each individual.

Generally, two types of selection methods are applied in EC, selection for reproduction and selection for survival [3] (see also Fig. 1.9). The first paradigm determines how to distribute the opportunity to reproduce among the individuals of the population; the latter paradigm determines how to administer the lifespan of different individuals for favoring the survival of promising individuals. Different EAs apply different combinations and implementations of these two selection criteria.

DE does not use the "selection for reproduction" mechanism, i.e., no individual is favored for reproduction compared to others [59]. In other words, it can be said that in DE each individual gets an opportunity to spawn its own offspring by mating with other individuals. Below we explain how the auxiliary parents (to whom the principal parent mates for breeding) are chosen.

Each individual in the current generation, irrespective of its fitness, becomes the principal parent for reproduction. Other individuals participating in reproduction, the auxiliary parents, are chosen randomly from the population. Specifically, for each individual $\vec{x}_i^G$, where G denotes the current generation, three other random individuals $\vec{x}_j^G$, $\vec{x}_k^G$, and $\vec{x}_l^G$ are selected from the population such that j, k and $l \in \{1, 2, \ldots, P\}$ and $i \neq j \neq k \neq l$. This way, a parent pool of four individuals is formed to breed an offspring.

DE applies selection pressure only when picking survivors. A knockout competition is played between each individual $\vec{x}_i^G$ and its offspring $\vec{u}_i^G$, and the winner is selected deterministically based on objective function values and promoted to the next generation. Thus, the survival criteria in DE can be described as follows:

$$\vec{x}_i^{G+1} = \begin{cases} \vec{u}_i^G & \text{if } f(\vec{u}_i^G) \leq f(\vec{x}_i^G) \\ \vec{x}_i^G & \text{otherwise,} \end{cases} \tag{2.4}$$

where $f(\cdot)$ indicates the objective function that is being optimized (minimized here). The one-to-one replacement strategy used in DE is different from what is generally observed in EAs. However, practicing this one-to-one selection mechanism enables DE to exercise elitism on its population. In fact, this scheme preserves not only the global best (the best individual of the population) but also the local best (the best individual encountered at any index). Note that the similar elitism is also used in PSO as described in Sect. 1.3.

On the other hand, using one-to-one survivor selection criteria, DE ignores many promising individuals, the exploitation of which could accelerate the search. Due to its positional elitism strategy, it discards an offspring which is better than most of the current population but worse than its parent. However, such rejected individuals could be useful to accelerate the search for the global optimum (see [53, 55] for details).

In any EA, a set of operators are used to alter the genetic code of current individuals to improve their fitness. The success and failure of the evolutionary system depend on carefully chosen operators. The most commonly used operators found in EAs are mutation and crossover. DE also makes use of these two operators. However, as we will see, DE maintains its uniqueness in using these operators.

DE derived its name from the mutation operator it applies to mutate its individual. The auxiliary parents selected for reproduction are engaged to generate the mutated individual. The scaled difference between two of the auxiliary parents are added to the third auxiliary parent to create the mutated individual. This operation is called "differential mutation" and generates the mutated individual $\vec{v}_i^G$, for the principal parent $\vec{x}_i^G$ according to the following equation:

$$\vec{v}_i^G = \vec{r}_1 + F \times (\vec{r}_2 - \vec{r}_3), \tag{2.5}$$

where F, commonly known as scaling factor or amplification factor, is a positive real number. Usually $\vec{r}_1$, $\vec{r}_2$, and $\vec{r}_3$ are randomly chosen. The suggested range for F is (0,1), based on empirical study [59]. Some studies suggest that the value of F less than 0.3 or 0.4 is less reliable and has no use at all [25, 71]. However, there are examples where $F < 0.3$ proved to be the most suitable choice for optimization [12, 54]. Nevertheless, it may be deduced that very small values of F are atypical but not impractical.

To complement the differential mutation search strategy, DE then uses a crossover operation, often referred to as "discrete recombination," in which the mutated individual $\vec{v}_i^G$ is mated with the principal parent $\vec{x}_i^G$ and generates the offspring or "trial individual" $\vec{u}_i^G$. This crossover operation used in classic DE is also known as "binomial crossover" which is actually a slightly modified version of the uniform crossover operation. In uniform crossover, the trial individual is created by choosing genes from either of the parents with uniform random probability C_r. The DE version of uniform crossover, i.e., the binomial crossover, uses the same strategy with the exception that at least one gene is inherited from the mutated individual $\vec{v}_i^G$.

Formally, the genes of $\vec{u}_i^G$ are inherited from $\vec{x}_i^G$ and $\vec{v}_i^G$, determined by the parameter crossover probability C_r, as follows:

$$u_{i,t}^G = \begin{cases} v_{i,t}^G & \text{if } r(t) \leq C_r \text{ or } t = rn(i) \\ x_{i,t}^G & \text{if } r(t) > C_r \text{ and } t \neq rn(i), \end{cases} \tag{2.6}$$

Algorithm 2.1 k-means algorithm

Assign each data point $\mathbf{x}_j$ to cluster $(j = 1, 2, \ldots, k)$ at random.
update :=TRUE
while *update* =TRUE **do** ▷ Terminates when there are no changes to the cluster allocations.
  update :=FALSE
  Calculate the centroids $\mathbf{v}_j (j = 1, 2, \ldots, k)$ of each cluster based on the data assignment
  Reallocate each $\mathbf{x}_i$ to the cluster with the closest center by finding the distance between each $\mathbf{x}_i$ and each $\mathbf{v}_j (j = 1, 2, \ldots, k)$.
  if any $\mathbf{x}_i$ changes its previous cluster **then**
    update :=TRUE
  end if
end while

where $(t = 1, 2, \ldots, N)$ denotes the tth element of individual vectors. $r(t) \in [0, 1]$ is the tth evaluation of a uniform random number generator and $rn(i) \in \{1, 2, \ldots, N\}$ is a randomly chosen index which ensures that $\vec{u}_i^G$ gets at least one element from $\vec{v}_i^G$. Originally, it was suggested that the value of C_r be chosen from [0,1]; later it was suggested that $0 \leq C_r \leq 0.2$ be used for decomposable functions and $0.9 \leq C_r \leq 1$ be used for indecomposable functions [59]. However, many different studies found that other settings of $C_r \in [0, 1]$ can also be effective for optimization [39, 52].

2.2 Machine Learning Techniques

2.2.1 k-Means Algorithm

The k-means algorithm is a non-hierarchical clustering method that uses the average value of clusters to classify all data into k clusters.

Given n data points $\mathbf{x}_j (j = 1, \ldots, n)$, the goal of clustering is to find the cluster X_i that minimizes the following function:

$$\text{Clustering\_error} = \sum_{i=1}^{k} \sum_{\mathbf{x} \in X_i} \|\mathbf{x} - \mathbf{v}_i\|, \tag{2.7}$$

where different clusters are disjoint $(X_i \cap X_j = \phi, i \neq j)$ and the union of all clusters covers the entire set of data points, and where $\| \cdot \|$ indicates Euclidean distance. The center of cluster X_i (normally taken as the average of the elements of the assigned data) is $\mathbf{v}_i$. In other words,

$$\mathbf{v}_i = \frac{1}{\mid X_i \mid} \sum_{\mathbf{x} \in X_i} \mathbf{x}, \tag{2.8}$$

and X_i is referred to as the centroid.

The k-means algorithm works as shown in Algorithm 2.1.

1. Assign each data point $\mathbf{x}_j$ to cluster $(1, 2, \ldots, k)$ at random,
2. Calculate the centroids $\mathbf{v}_j (j = 1, 2, \ldots, k)$ of each cluster based on the data assignment,
3. Reallocate each $\mathbf{x}_i$ to the cluster with the closest center by finding the distance between each $\mathbf{x}_i$ and each $\mathbf{v}_j (j = 1, 2, \ldots, k)$,
4. This processing terminates when there are no changes to the cluster allocations. When changes have occurred, this processing is repeated, recalculating the centroids $\mathbf{v}_j$ based on the new cluster allocations.

This algorithm takes the number of clusters k as a parameter, and so another method is required to select the optimal number of clusters.

The k means algorithm yields a local optimum to the target function (i.e., Eq. (2.7)), but this is not necessarily the global minimum. The clustering results are influenced by the initial clusters that are assigned at random in the beginning.

2.2.2 SVM

The support vector machine (SVM) [15, 69] is a supervised learning method that is widely used in classification and regression analysis because of its high generalization performance. We first explain soft margin linear SVM as a simple example. Multiple feature vectors, $\mathbf{x}_1, \ldots, \mathbf{x}_m$, and corresponding labels, $y_1, \ldots, y_m (= +1 \text{ or} -1)$, are provided. Positive example data are $y_i = +1$, and negative example data are $y_i = -1$.

The objective of learning is to find the separating hyperplane that separates the two classes (i.e., $y_i = +1 \text{ or} -1$) and generates a classifier of unknown data. As shown in Fig. 2.4, soft margin linear SVM finds a hyperplane that separates two classes. There are multiple hyperplane candidates, but the most desirable solution is one that can best categorize unknown data, namely one with high generalization performance. Therefore, the hyperplane that maximizes the margins from individual training data is sought as the optimum solution. The concept of soft margin was introduced to appropriately handle training data that was categorized in the incorrect region as a result of measurement error (as one example). Incorrect data are penalized and the sum of the penalties is minimized. Therefore, soft margin linear SVM learns to find a hyperplane that offers a desirable trade-off between margin maximization and penalty minimization.

A detailed discussion about linear SVM with reference to [20, 31] is as follows. Let $\mathbf{X} = \{(\mathbf{x}_i, l_i) \mid \mathbf{x}_i \in \mathbb{R}^d,\ l_i \in \{+1, -1\}\}_{i=1}^N$ be a d-dimensional dataset containing two classes and N instances. For a linearly separable dataset,

$$\exists \mathbf{w} \quad \text{s.t.} \quad l_i(\mathbf{w}^\top \mathbf{x}_i + b) \geq 1, \quad \forall i = 1, 2, \ldots, N. \tag{2.9}$$

An optimal separating plane lies mid-way between the two planes $P_1 : \mathbf{w}^\top \mathbf{x} + b = -1$ and $P_2 : \mathbf{w}^\top \mathbf{x} + b = 1$. The margin m of the optimal separating plane is found by taking two support vectors $\mathbf{x}_1$ and $\mathbf{x}_2$ on the plane P_1 and P_2, respectively, and is

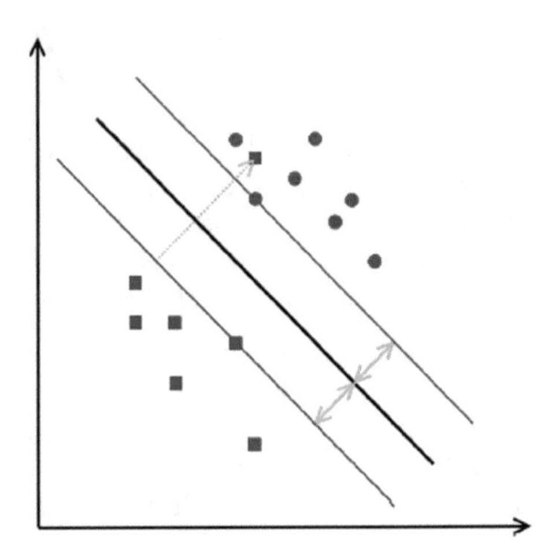

Fig. 2.4 Example of soft margin linear SVM

given by

$$m = \left(\frac{\mathbf{w}}{||\mathbf{w}||}\right)^{\top}(\mathbf{x}_2 - \mathbf{x}_1) = \frac{2}{||\mathbf{w}||}. \tag{2.10}$$

The norm of the weight, i.e., $||\mathbf{w}||$, has to be minimized in order to maximize the margin m. Therefore, training an SVM becomes the following optimization problem:

$$\begin{aligned} \underset{\mathbf{w},b}{\text{Minimize}} \quad & \frac{1}{2}\mathbf{w}^{\top}\mathbf{w} \\ \text{Subject to} \quad & l_i(\mathbf{w}^{\top}\mathbf{x}_i + b) \geq 1, \quad \forall i = 1, 2, \ldots, N. \end{aligned} \tag{2.11}$$

In case of a non-separable dataset, slack variables $\xi_i \geq 0$ are introduced. The introduction of slack variable ξ_i allows us to consider a margin of error for wrongly classified data points. Then using a parameter $C > 0$ called regularization constant of SVM, the above optimization problem is given as follows:

$$\begin{aligned} \underset{\mathbf{w},b}{\text{Minimize}} \quad & \frac{1}{2}\mathbf{w}^{\top}\mathbf{w} + C\sum_{i=1}^{N}\xi_i \\ \text{Subject to} \quad & l_i(\mathbf{w}^{\top}\mathbf{x}_i + b) \geq 1 - \xi_i, \quad \forall i = 1, 2, \ldots, N. \end{aligned} \tag{2.12}$$

Training the SVM requires solving the optimization problem represented by Eq. (2.12) in order to find the weight $\mathbf{w}$ and the bias b. The computational complexity of the training phase is $O(N \cdot d)$ which scales linearly with the size of the dataset. On the other hand, testing a sample $\mathbf{x}_{test}$ requires evaluating

$$sgn(\mathbf{w}^{\top}\mathbf{x}_{test} + b). \tag{2.13}$$

The computational complexity of the testing phase is $O(d)$ which makes a linear SVM very fast.

In case of non-separable dataset, the classification accuracy can be increased by transforming the decision boundary into a linear decision boundary by some mapping function, $\phi : \mathbf{x} \mapsto \phi(\mathbf{x})$. Unfortunately, in many cases, the linear decision boundary is only obtained in a very high-dimensional feature space or even an infinite dimensional feature space. As a result, the inner product in Eq. (2.12), may be computationally demanding or even impossible. To solve this problem, special functions known as kernels, $k(\mathbf{x}_i, \mathbf{x}_j)$ are used which represent the inner product of vectors implicitly mapped into higher dimension, i.e., $k(\mathbf{x}_i, \mathbf{x}_j) = \phi(\mathbf{x}_i)^\top \phi(\mathbf{x}_j)$.

For examples, the following kernels are frequently used:

- Liner kernel

$$k(\mathbf{x}_i, \mathbf{x}_j) = \mathbf{x}_i^\top \mathbf{x}_j \tag{2.14}$$

- Polynomial kernel

$$k(\mathbf{x}_i, \mathbf{x}_j) = (a \times \mathbf{x}_i^\top \mathbf{x}_j + b)^c \tag{2.15}$$

- RBF (Radial basis function) kernel or Gaussian kernel

$$k(\mathbf{x}_i, \mathbf{x}_j) = \exp(-\sigma \|\mathbf{x}_i - \mathbf{x}_j\|^2) \;\; (\sigma > 0) \tag{2.16}$$

The linear algorithm which is based on Eq. (2.12) has to be adjusted so that kernel functions can be used. This adjustment is known as kernel trick. In order to use the kernel trick, the dual Lagrangian form of Eq. (2.12) is used. Considering the mapping of the feature vectors, the dual Lagrangian is

$$\begin{aligned} &\underset{\alpha_i}{\text{Maximize}} \quad L(\alpha_i) = \sum_{i=1}^{N} \alpha_i - \frac{1}{2} \sum_{i,j=1}^{N} \alpha_i \alpha_j l_i l_j \phi(\mathbf{x}_i)^T \phi(\mathbf{x}_j) \\ &\text{Subject to} \quad 0 \leq \alpha_i \leq C, \quad \sum_{i=1}^{N} \alpha_i l_i = 0, \quad \forall i = 1, 2, \ldots, N, \end{aligned} \tag{2.17}$$

where α_i is the Lagrange multiplier. Equation (2.17) is a standard quadratic optimization problem in $\alpha_i \geq 0$. This dual form is used for nonlinear SVM [13].

Training the SVM requires solving the optimization problem represented by Eq. (2.17) in order to find the Lagrangian multipliers, α_i. An interesting thing to note about this equation is that the inner product of all the pairs of points $\mathbf{x}_i$ and $\mathbf{x}_j$ are used. This requires the calculation of $N \times N$ Gram matrix.[1] As a result of which, the computational complexity of training is $O(N^2 \cdot d)$ which is unacceptable for large dataset. Thus, there is trade-off between accuracy and speed with nonlinear SVM. On the other hand, testing a sample $\mathbf{x}_{test}$ requires evaluating

[1]Given a set of vectors $\mathbf{v}_1, \ldots, \mathbf{v}_n$, its Gram matrix G is an $n \times n$ matrix, whose element is an inner product of two vectors, i.e., $G_{ij} = \mathbf{v}_i \cdot \mathbf{v}_j$.

$$\operatorname{sgn}\left(\sum_{i=1}^{N}\alpha_i l_i \phi(\mathbf{x}_i)^{\top}\phi(\mathbf{x}_{test}) + b\right). \tag{2.18}$$

Therefore, the computational complexity of the testing phase is $O(N \cdot d)$ and scales linearly with the size of the dataset.

2.2.3 RVM: Relevance Vector Machine

This section describes the sparse Bayesian kernel method, i.e., Relevance Vector Machine (RVM). RVM is a generalized linear model of identical functional form to SVM and solves the above disadvantages. In Sect. 4.4, we show how RVM is employed to find suitable basis functions in GP.

SVM has the following disadvantages [68]:

- Predictions are not probabilistic. Because SVM outputs a point estimate.
- The requisite number of kernel functions grows steeply with the size of the training set.
- SVM is necessary to estimate the error/margin trade-off parameter.

Given a train set of input-target pairs $\{\mathbf{x}_n, t_n\}_{n=1}^{N}$, we assume that the target value is generated by the output ground-truth function $y(\mathbf{x}_\mathrm{n})$ with additive noise ε_n. The ground-truth function is approximated by a linear combination of a set of nonlinear basis (feature) functions $\sum_{i=1}^{M} w_i \phi_i(\mathbf{x}_n)$:

$$\begin{aligned} t_n &= y(\mathbf{x}_\mathrm{n}) + \varepsilon_n \\ &= \textstyle\sum_{i=1}^{M} w_i \phi_i(\mathbf{x}_n) + \varepsilon_n \\ &= \mathbf{w}^\mathrm{T}\phi(\mathbf{x}_n) + \varepsilon_n, \end{aligned} \tag{2.19}$$

where ε_n is assumed to be independently sampled from a zero-mean Gaussian distribution with precision β (variance $\sigma^2 = \beta^{-1}$). The functions $\phi_i(\mathbf{x})$ are the fixed nonlinear basis functions, which are considered known to us.

Probabilistically, we assume a conditional distribution for the real-valued target variable t, conditioned on input variable $\mathbf{x}$, undetermined linear coefficient $\mathbf{w}$, and the precision parameter β of the Gaussian noise. The pdf (probability density function) of t takes the form:

$$p(t \mid \mathbf{x}, \mathbf{w}, \beta) = \mathcal{N}\left(t \mid \mathbf{w}^\mathrm{T}\phi(\mathbf{x}), \beta^{-1}\mathbf{I}\right), \tag{2.20}$$

where Gaussian distribution is noted as $\mathcal{N}$. For a D-dimensional random variable $\mathbf{z}$, the pdf of the multivariate Gaussian takes the form:

$$\mathcal{N}(\mathbf{z} \mid \boldsymbol{\mu}, \mathbf{C}) = \frac{1}{(2\pi)^{\frac{D}{2}}}\frac{1}{|\mathbf{C}|^{\frac{1}{2}}}\exp\left\{-\frac{1}{2}(\mathbf{x}-\boldsymbol{\mu})^\mathrm{T}\mathbf{C}^{-1}(\mathbf{x}-\boldsymbol{\mu})\right\}, \tag{2.21}$$

where $\boldsymbol{\mu}$ is a D-dimensional vector known as the expectation or mean, $\mathbf{C}$ is a $D \times D$ matrix known as the covariance matrix, and $|\mathbf{C}|$ denotes the determinant of $\mathbf{C}$.

Since we have assumed the independent noise on target, the likelihood of the given data can be simply written in the form of product of each single pdf on t_n:

$$p\left(\mathbf{t} \mid \mathbf{X}, \mathbf{w}, \beta\right) = \prod_{n=1}^{N} p(t_n|\mathbf{x}_n, \mathbf{w}, \beta). \tag{2.22}$$

The maximum likelihood estimation (MLE) is to estimate the unknown parameters $\mathbf{w}$ and β by maximizing the likelihood function. Here, we implicitly find the derivate of the likelihood and set it to zero to obtain the following results of MLE:

$$\mathbf{w}_{\mathrm{ML}} = \left(\boldsymbol{\Phi}^{\mathrm{T}}\boldsymbol{\Phi}\right)^{-1}\boldsymbol{\Phi}^{\mathrm{T}}\mathbf{t}, \tag{2.23}$$

$$\beta_{ML}^{-1} = \frac{1}{N}\sum_{n=1}^{N}\{t_n - \mathbf{w}_{\mathrm{ML}}^{\mathrm{T}}\phi(\mathbf{x}_n)\}, \tag{2.24}$$

where $\boldsymbol{\Phi}$ is a $N \times M$ matrix, which is known as design matrix, and whose elements are given by $\Phi_{ni} = \phi_i\left(\mathbf{x}_n\right)$, i.e.,

$$\boldsymbol{\Phi} = \begin{pmatrix} \phi_1(\mathbf{x}_1) & \phi_2(\mathbf{x}_1) & \cdots & \phi_M(\mathbf{x}_1) \\ \phi_1(\mathbf{x}_2) & \phi_2(\mathbf{x}_2) & & \phi_M(\mathbf{x}_2) \\ \vdots & & \ddots & \vdots \\ \phi_1(\mathbf{x}_N) & \phi_2(\mathbf{x}_N) & \cdots & \phi_M(\mathbf{x}_N) \end{pmatrix}. \tag{2.25}$$

Our discussion follows the Bayesian treatment of linear regression by introducing a prior distribution over the linear coefficients $\mathbf{w}$. Temporarily, we view the noise precision β as a known constant. Due to the reason that Eq. (2.22) is the exponential of a quadratic function of $\mathbf{w}$, it will be analytically convenient to introduce a Gaussian prior distribution for $\mathbf{w}$:

$$p\left(\mathbf{w} \mid \alpha\right) = \mathcal{N}(\mathbf{w}|\mathbf{0}, \alpha^{-1}\mathbf{I}). \tag{2.26}$$

For the moment, we have introduced a single shared precision parameter α, which is the assumption of a normal Bayesian linear regression instead of RVM. However, it is necessary to explain the general approach we used in Bayesian parameter estimation first. After we have assumed that $\mathbf{w}$ is a zero-mean Gaussian, the corresponding posterior distribution over $\mathbf{w}$ should be positively correlated to the product of the likelihood function and the prior distribution. Our choice of a conjugate prior to the quadratic likelihood results in an antically clear form of the posterior, which is still Gaussian. The evaluation of this distribution should be performed by the mathematical method known as completing the square in the exponential terms, and then finding the normalization coefficient using the standard result for a normalized Gaussian. Since the Bayes' theorem for Gaussian variables will be used once again in the discussion of RVM, here we present the general results, which are referenced from [4].

Assume that a Gaussian marginal distribution $p(\mathbf{z}_1)$ over $\mathbf{z}_1$, and a Gaussian conditional distribution $p(\mathbf{z}_2|\mathbf{z}_1)$ over $\mathbf{z}_2$ are given. We concern the marginal distribution $p(\mathbf{z}_2)$ over $\mathbf{z}_2$, and the conditional distribution $p(\mathbf{z}_1|\mathbf{z}_2)$ over $\mathbf{z}_1$.

The marginal and conditional distribution can be written in [4]:

$$p\left(\mathbf{z}_1\right) = \mathcal{N}(\mathbf{z}_1|\boldsymbol{\mu}, \boldsymbol{\Lambda}^{-1}), \tag{2.27}$$

$$p\left(\mathbf{z}_2|\mathbf{z}_1\right) = \mathcal{N}(\mathbf{z}_2|\mathbf{A}\mathbf{z}_1+\mathbf{b}, \mathbf{L}^{-1}), \tag{2.28}$$

where $\boldsymbol{\mu}$,$\mathbf{A}$ and $\mathbf{b}$ are parameters controlling the expectations (means); $\boldsymbol{\Lambda}$ and $\mathbf{L}$ are precision matrices, which are inverses of the covariance matrices. If $\mathbf{z}_1$ is a M-dimensional random variable, and $\mathbf{z}_2$ is a D-dimensional one, then the matrix $\mathbf{\Lambda}$ results in the size of $D \times M$.

The joint distribution is fundamental to the Bayes' theory. To find it, we consider it as the distribution over the jointed vector $\mathbf{z}$, which is defined by [4]:

$$\mathbf{z} = \begin{pmatrix} \mathbf{z}_1 \\ \mathbf{z}_2 \end{pmatrix}. \tag{2.29}$$

We then consider the logarithm of the joint distribution, which should be given by the summation of the logarithms of the marginal and conditional distribution, according to Bayes' theory [4]:

$$\begin{aligned} \ln p\left(\mathbf{z}\right) &= \ln p\left(\mathbf{z}_1\right) + \ln p\left(\mathbf{z}_1 \mid \mathbf{z}_2\right) \\ &= -\tfrac{1}{2}\left(\mathbf{z}_1 - \boldsymbol{\mu}\right)^{\mathrm{T}} \boldsymbol{\Lambda} \left(\mathbf{z}_1 - \boldsymbol{\mu}\right) - \tfrac{1}{2}\left(\mathbf{z}_2 - \mathbf{A}\mathbf{z}_1 - \mathbf{b}\right)^{\mathrm{T}} \mathbf{L} \left(\mathbf{z}_2 - \mathbf{A}\mathbf{z}_1 - \mathbf{b}\right) + const. \end{aligned} \tag{2.30}$$

Here, $const$ denotes redundant terms unrelated to $\mathbf{z}_1$ and $\mathbf{z}_2$. Since the logarithm of $p\left(\mathbf{z}\right)$ is a quadratic function of $\mathbf{z}$, $p(\mathbf{z})$ should be a Gaussian distribution. We can thus find the precision matrix of this distribution by considering the second-order terms in Eq. (2.30), which are given by [4]:

$$-\frac{1}{2}\begin{pmatrix} \mathbf{z}_1 \\ \mathbf{z}_2 \end{pmatrix}^{\mathrm{T}} \begin{pmatrix} \boldsymbol{\Lambda}+\mathbf{A}^{\mathrm{T}}\mathbf{L}\mathbf{A} & -\mathbf{A}^{\mathrm{T}}\mathbf{L} \\ -\mathbf{L}\mathbf{A} & \mathbf{L} \end{pmatrix} \begin{pmatrix} \mathbf{z}_1 \\ \mathbf{z}_2 \end{pmatrix} = -\frac{1}{2}\mathbf{z}^{\mathrm{T}}\mathbf{R}\mathbf{z}. \tag{2.31}$$

Therefore, by comparing the pdf of the Gaussian (i.e., Eqs. (2.21) and (2.31), we are able to conclude that the distribution over jointed variable $\mathbf{z}$ has a covariance matrix given by the inverse of the precision matrix [4]:

$$\mathbf{Cov}[\mathbf{z}] = \mathbf{R}^{-1} = \begin{pmatrix} \boldsymbol{\Lambda}^{-1} & \boldsymbol{\Lambda}^{-1}\mathbf{A}^{\mathrm{T}} \\ \mathbf{A}\boldsymbol{\Lambda}^{-1} & \mathbf{L}^{-1}+\mathbf{A}\boldsymbol{\Lambda}^{-1}\mathbf{A}^{\mathrm{T}} \end{pmatrix}. \tag{2.32}$$

The expectation vector of the Gaussian distribution over $\mathbf{z}$ thus can be determined by analyzing the linear terms in Eq. (2.30), which are given by [4]:

$$\mathbf{z}_1^{\mathrm{T}}\boldsymbol{\Lambda}\boldsymbol{\mu}-\mathbf{z}_1^{\mathrm{T}}\mathbf{A}^{\mathrm{T}}\mathbf{L}\mathbf{b}+\mathbf{z}_2^{\mathrm{T}}\mathbf{L}\mathbf{b}=\begin{pmatrix}\mathbf{z}_1\\ \mathbf{z}_2\end{pmatrix}^{\mathrm{T}}\begin{pmatrix}\boldsymbol{\Lambda}\boldsymbol{\mu}-\mathbf{A}^{\mathrm{T}}\mathbf{L}\mathbf{b}\\ \mathbf{L}\mathbf{b}\end{pmatrix}. \tag{2.33}$$

Similarly, the expectation vector of $\mathbf{z}$ can be found by comparing Eq. (2.21)–(2.31), since the expectation is only related to linear terms in the exponentials of the pdf expression. The result is given by [4]:

$$\begin{aligned}\mathrm{E}[\mathbf{z}] &= \mathbf{R}^{-1}\begin{pmatrix}\boldsymbol{\Lambda}\boldsymbol{\mu}-\mathbf{A}^{\mathrm{T}}\mathbf{L}\mathbf{b}\\ \mathbf{L}\mathbf{b}\end{pmatrix}\\ &= \begin{pmatrix}\boldsymbol{\mu}\\ \mathbf{A}\boldsymbol{\mu}+\mathbf{b}\end{pmatrix}.\end{aligned} \tag{2.34}$$

Until now, we have found the covariance matrix Eq. (2.32) and expectation vector Eq. (2.34) of the Gaussian distribution over the jointed variable $\mathbf{z}$. Next, the marginal distribution of $\mathbf{z}_2$ is to be determined. For a marginal distribution, the expectation and covariance can be simply expressed in terms of the partitioned covariance matrix of the joint distribution according to [4]:

$$\mathrm{E}[\mathbf{z}_2]=\mathbf{A}\boldsymbol{\mu}+\mathbf{b}, \tag{2.35}$$

$$\mathbf{Cov}[\mathbf{z}_2]=\mathbf{L}^{-1}+\mathbf{A}\boldsymbol{\Lambda}^{-1}\mathbf{A}^{\mathrm{T}}. \tag{2.36}$$

For the conditional distribution, the results are as follows [4]:

$$\mathrm{E}[\mathbf{z}_1|\mathbf{z}_2]=\left(\boldsymbol{\Lambda}+\mathbf{A}^{\mathrm{T}}\mathbf{L}\mathbf{A}\right)^{-1}\left\{\mathbf{A}^{\mathrm{T}}\mathbf{L}\left(\mathbf{z}_2-\mathbf{b}\right)+\boldsymbol{\Lambda}\boldsymbol{\mu}\right\}, \tag{2.37}$$

$$\mathbf{Cov}[\mathbf{z}_1|\mathbf{z}_2]=\left(\boldsymbol{\Lambda}+\mathbf{A}^{\mathrm{T}}\mathbf{L}\mathbf{A}\right)^{-1}. \tag{2.38}$$

Making use of the results from Eqs. (2.37) and (2.38), together with Eqs. (2.22) and (2.26), we obtain the posterior distribution over $\mathbf{w}$:

$$p(\mathbf{w}\mid\mathbf{t})=\mathcal{N}(\mathbf{w}|\mathbf{m},\Sigma), \tag{2.39}$$

with the posterior mean:

$$\mathbf{m}=\beta\boldsymbol{\Sigma}\boldsymbol{\Phi}^{\mathrm{T}}\mathbf{t}, \tag{2.40}$$

and the posterior covariance:

$$\boldsymbol{\Sigma}=\left(\alpha^{-1}\mathbf{I}+\beta\boldsymbol{\Phi}^{\mathrm{T}}\boldsymbol{\Phi}\right)^{-1}. \tag{2.41}$$

Maximization of this posterior distribution will determine $\mathbf{w}$ by finding the most probable value of $\mathbf{w}$ given data. This technique is called maximum posterior (MAP):

$$\mathbf{w}_{MAP}=\mathbf{m}. \tag{2.42}$$

Different from the assumption that only one α is shared for all weight parameter w_i, RVM constrains the parameters by introducing a separate hyperparameter α_i for each of the weight parameter w_i. Therefore, it results in the generalized form of Eq. (2.26) in the form [4]:

$$p\,(\mathbf{w}\,|\,) = \prod_{i=1}^{M} \mathcal{N}(w_i|0,\alpha_i^{-1}), \tag{2.43}$$

where α_i is treated as the precision (inverse variance) of the corresponding parameter w_i, and denotes $(\alpha_1, \ldots, \alpha_M)^{\mathrm{T}}$. The introduction of an individual hyperparameter for every weight is the mainpoint of the model. If we estimate these hyperparameters, most of them result in infinity, and the corresponding weights w_i take posterior distributions with infinitely high precision (infinitely low variance) to zero. Each basis function associated with these parameters therefore is multiplied by a zero weight and contributes nothing to the model, thus can be automatically pruned out from the model.

To estimate these unknown hyperparameters, we first obtain the posterior distribution over weights by using Bayes' theory:

$$p\,(\mathbf{w}\,|\,\mathbf{t}, \mathbf{X}, ,\beta) = \mathcal{N}(\mathbf{w}|\mathbf{m}, \Sigma), \tag{2.44}$$

with the posterior mean:

$$\mathbf{m} = \beta\boldsymbol{\Sigma}\boldsymbol{\Phi}^{\mathrm{T}}\mathbf{t}, \tag{2.45}$$

and posterior covariance:

$$\boldsymbol{\Sigma} = \left(\mathbf{A} + \beta\boldsymbol{\Phi}^{\mathrm{T}}\boldsymbol{\Phi}\right)^{-1}, \tag{2.46}$$

where $\mathbf{A} = \mathrm{diag}(\alpha_i)$, i.e., a square diagonal matrix with α_i on the main diagonal.

Next, the training of RVM requires to search for posterior hyperparameter to maximize the marginal likelihood function [4]:

$$p\,(\mathbf{t}\,|\,\mathbf{X}, ,\beta) = \int p\,(\mathbf{t}\,|\,\mathbf{X}, \mathbf{w},\beta)\; p(\mathbf{w}|)d\mathbf{w}. \tag{2.47}$$

This integration involves a convolution of two Gaussian functions, which can be analytically evaluated by completing the spare to give the logged marginal likelihood with [4]:

$$\ln\,(p\,(\mathbf{t}\,|\,\mathbf{X}, ,\beta)) = -\frac{1}{2}\left\{N\ln\,(2\pi) + \ln\,(|\mathbf{C}|) + \mathbf{t}^{\mathrm{T}}\mathbf{C}^{-1}\mathbf{t}\right\}, \tag{2.48}$$

where the N $\times$ N matrix **C** is defined by [4]:

$$\mathbf{C} = \beta^{-1}\mathbf{I} + \boldsymbol{\Phi}\mathbf{A}^{-1}\boldsymbol{\Phi}^{\mathrm{T}}. \tag{2.49}$$

Our objective is to maximize Eq. (2.48) w.r.t the hyperparameters and β. We take derivatives of the marginal likelihood w.r.t. each α_i, i.e.,

$$\frac{\partial}{\partial \alpha_i} \ln\left(p\left(\mathbf{t} \mid \mathbf{X},,\beta\right)\right)=\frac{1}{2\alpha_i}-\frac{1}{2}\Sigma_{ii}-\frac{1}{2}m_i^2, \tag{2.50}$$

where m_i is the ith component of the posterior mean $\mathbf{m}$ defined by Eq. (2.45), Σ_{ii} is the ith diagonal component of the posterior covariance $\mathbf{\Sigma}$ given by Eq. (2.46).

Setting this derivative to zero, we obtain the following equation:

$$\alpha_i=\frac{1-\alpha_i\Sigma_{ii}}{m_i^2}=\frac{\gamma_i}{m_i^2}, \tag{2.51}$$

where we have defined as follows:

$$\gamma_i= 1-\alpha_i\Sigma_{ii}. \tag{2.52}$$

Similarly, for β we see that:

$$\frac{\partial}{\partial \beta} \ln\left(p\left(\mathbf{t} \mid \mathbf{X},,\beta\right)\right)=\frac{1}{2}\left(\frac{N}{\beta}-\|\mathbf{t}-\mathbf{\Phi}\boldsymbol{m}\|^2-\mathrm{Tr}[\mathbf{\Sigma}\mathbf{\Phi}^{\mathrm{T}}\mathbf{\Phi}]\right). \tag{2.53}$$

Using Eq. (2.46), we can rewrite the argument of the trace operator as:

$$\begin{aligned}\mathbf{\Sigma}\mathbf{\Phi}^{\mathrm{T}}\mathbf{\Phi} &= \mathbf{\Sigma}\mathbf{\Phi}^{\mathrm{T}}\mathbf{\Phi}+\beta^{-1}\mathbf{\Sigma}\boldsymbol{A}-\beta^{-1}\mathbf{\Sigma}\boldsymbol{A}\\ &= \mathbf{\Sigma}\left(\mathbf{\Phi}^{\mathrm{T}}\mathbf{\Phi}\beta+\mathbf{A}\right)\beta^{-1}-\beta^{-1}\mathbf{\Sigma}\boldsymbol{A}\\ &= \left(\mathbf{\Phi}^{\mathrm{T}}\mathbf{\Phi}\beta+\mathbf{A}\right)^{-1}\left(\mathbf{\Phi}^{\mathrm{T}}\mathbf{\Phi}\beta+\mathbf{A}\right)\beta^{-1}-\beta^{-1}\mathbf{\Sigma}\boldsymbol{A}\\ &= (\mathbf{I}-\mathbf{A}\mathbf{\Sigma})\,\beta^{-1}.\end{aligned} \tag{2.54}$$

Now setting Eq. (2.53) to zero, and use the result in Eq. (2.54), we obtain:

$$\beta_i=\frac{\|\mathbf{t}-\mathbf{\Phi}\mathbf{m}\|^2}{\mathrm{Tr}(\mathbf{I}-\mathbf{A}\mathbf{\Sigma})}=\frac{\|\mathbf{t}-\mathbf{\Phi}\mathbf{m}\|^2}{N-\sum_i\gamma_i}. \tag{2.55}$$

Note that Eqs. (2.51) and (2.53) are implicit solutions because γ_i and m_i^2 depends on $\boldsymbol{\alpha}$. We therefore adopt an iterative procedure by choosing initial values for $\boldsymbol{\alpha}$, β, evaluating the mean and covariance of the posterior using Eqs. (2.45) and (2.46), respectively, and then alternately re-estimating (1) the hyperparameters using Eqs. (2.51) and (2.55), and (2) the posterior mean and covariance, using Eqs. (2.45) and (2.46), until a suitable convergence criterion is satisfied [4].

2.2.4 k-Nearest Neighbor Classifier

The k-nearest neighbor (k-NN) classifier [16], an extension of nearest neighbor classifier [1], has long been used in pattern recognition and machine learning for supervised classification tasks. The basic approach involves storing all the training instances;

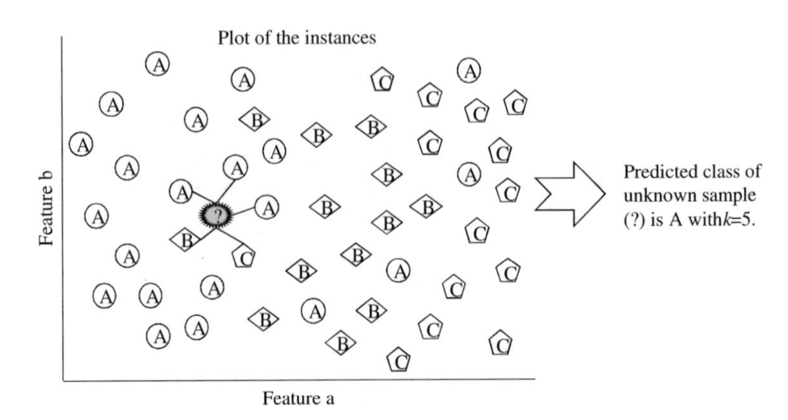

Fig. 2.5 An example of class prediction by the kNN classifier

when a test instance is presented, retrieving k training instances that are nearest to this test instance and prediction of the label of the test instance by majority voting. The distance between two instances $\mathbf{x} = (x_1, x_2, \ldots, x_n)$ and $\mathbf{y} = (y_1, y_2, \ldots, y_n)$ is calculated as follows:

$$d(\mathbf{x}, \mathbf{y}) = \sqrt{\sum_{i=1}^{n} w_i (x_i - y_i)^2}, \tag{2.56}$$

where n is the number of features in the data set and w_i is the weight of feature i. When w_i is set to 1, the distance between two instances becomes Euclidean distance. To avoid a tie, the value of k should be an odd number (and of course, k is smaller than the number of training samples) for binary classification. An example of the k-NN classifier is given in Fig. 2.5.

k-NN is very easy to implement and can provide good classification accuracy if the features are chosen and weighted carefully in computation of distance.

When the instances are of variable lengths or discrete time series data, such as the health checkup and lifestyle data of an organization, distance between two instances cannot be calculated by directly applying the above Eq. (2.56). In [57], two methods are proposed, namely the aggregation method and the sliding window method, to calculate the distance between two variable length instances and determine the nearest neighbors.

In the aggregation method, the length of each instance is made equal by aggregating the values of a feature across various observations into a single value, and then the distance between two equal length instances is calculated. The aggregation method that is to be applied for a feature depends on the data type of that feature. If a feature is numeric, a function that returns a numeric value should be used; if the feature is nominal, a function that returns a nominal value should be used. Some examples of aggregation functions for numeric features are *average*, *max*, and *min*;

one example of an aggregation function for nominal features is *mode* that returns the most frequent value.

In the sliding window method, distances between two instances are calculated at different window positions of the larger instance, and the window having the minimum distance is taken as the best matching window. The window is moved one data point at a time where each data point consists of either the value of a single observation, such as the output signal amplitude of a sensor at time t or the values of a number of features under single observation, such as the health checkup and lifestyle data of a person in a year. If the number of data points in two instances is respectively m and n $(n \leq m)$, the number of sliding windows to consider to find the minimum distance is $m - n + 1$. Let us give an example of the sliding window method to find the minimum distance in two instances consisting of the following data points: $X_1X_2X_3X_4X_5$ and $Y_1Y_2Y_3$. To find the minimum distance, three sliding windows are considered and distances between $X_1X_2X_3$ and $Y_1Y_2Y_3$; $X_2X_3X_4$ and $Y_1Y_2Y_3$; and $X_3X_4X_5$ and $Y_1Y_2Y_3$ are calculated.

2.2.5 *Transfer Learning*

Transfer learning [56] is a framework in machine learning where data on related problems and derived knowledge are used to effectively and efficiently solve the target problem. The sender and the receiver of knowledge to be transferred are called the source domain and target domain, respectively.

Humans learn various things from transfer learning. For instance, when someone who can play the piano starts to learn the electronic organ at the same time as someone who cannot play the piano, the former can learn to play the electronic organ better and in a shorter amount time than the latter. In the framework of transfer learning, the piano is the source domain, the electronic organ is the target domain, and proficiency in playing the piano assists in the learning of the electronic organ. Transfer learning is applied in various fields, including natural language processing, voice recognition, and image processing.

Figure 2.6 shows a rough flow of transfer learning. In this example, the source task concerns training on female speech and the target task is to recognize speech from males. Learning is carried out using data and knowledge associated with problems in the source and target domains to ultimately answer problems in the target domain efficiently and with high precision. Usually, it is assumed that the source and the target domains have some structural relationship.

Transfer learning is very effective when there is little training data in the target domain, but substantial data in the source domain. Moreover, transferring knowledge from a domain that is highly similar to the target domain results in more efficient learning. In contrast, the transfer of knowledge from a source domain with low similarity results in a decrease in learning performance, which is called negative transfer.

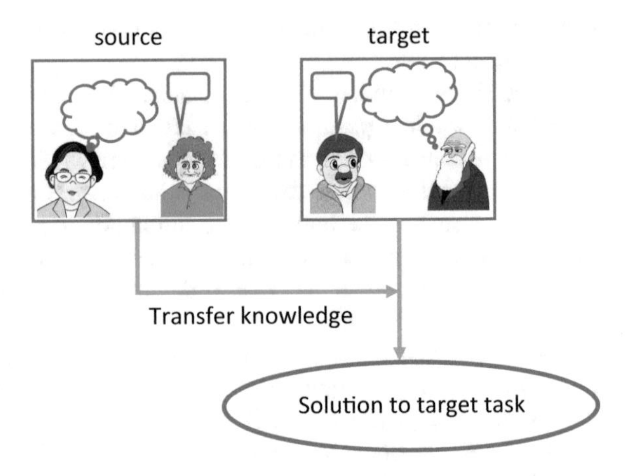

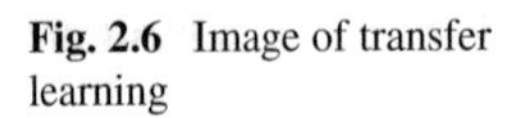
Fig. 2.6 Image of transfer learning

In transfer learning, the maximum limit of learning performance in the training domain is normally limited by the learning performance in the source domain. In other words, a higher learning precision in the source domain results in a better chance of improving learning efficiency in the target domain.

An important problem in transfer learning is determining what knowledge to transfer. If the data from the source domain is necessary, the most obvious approach is to appropriately map the data and use it when learning in the target domain. On the other hand, transferring feature values or parameters that exist both in the source and target domains is also possible. What knowledge can be transferred and what knowledge successfully works depends on each domain, and determining which to apply is difficult.

2.2.6 Bagging and Boosting

Boosting and bagging are general resampling methods for improving the performance of any learning algorithm [22]. Both methods work by repeatedly running a given weak learning algorithm on various distributions over the training data. They rely on a resampling technique to obtain a different training set for each classifier. Boosting can theoretically be used to significantly reduce the error of any weak learning algorithm which need only be a little bit better than random guessing. The resulting classifier has been shown to be more accurate than any of the individual classifiers making up the ensemble [48].

The data for learning systems are supposed to consist of attribute-value vectors or instances. Both boosting and bagging manipulate the training data in order to generate different classifiers. Following [17, 60], we shall explain the basic ideas of bagging and boosting below.

2.2.6.1 Bagging

Bagging produces replicate training sets by sampling with replacement from the training instances. For each trial $t = 1, 2, \ldots, T$, a training set of size N is sampled with replacement from the original instances. This training set is the same size as the original data, but some instances may not appear in it while others appear more than once. The learning system generates a classifier C^t from the sample and the final classifier C^* is formed by aggregating the T classifiers from these trials. To classify an instance x, a vote for class k is recorded by every classifier for which $C^t(x) = k$ and C^* is then the class with the most votes (ties being resolved arbitrarily).

2.2.6.2 Boosting

Boosting uses all instances at each repetition, but maintains a weight for each instance in the training set that reflects its importance adjusting the weights causes the learner to focus on different instances and leads to different classifiers. When voting to form a composite classifier, each component classifier has the same vote in bagging, whereas boosting assigns different voting strengths to component classifiers on the basis of their accuracy.

Let w_x^t denote the weight of instance x at trial t. Initially, $w_x^1 = 1/N$ for every x. At each trial $t = 1, 2, \ldots, T$, a classifier C^t is constructed from the given instances under the distribution w^t, as if the weight w_x^t of instance x reflects its probability of occurrence. The error ϵ^t of this classifier is also measured with respect to the weights and consists of the sum of the weights of the instances that it misclassified. If ϵ^t is greater than 0.5, the trials are terminated, and T is altered to $t - 1$. Conversely, if C^t correctly classifies all instances so that ϵ^t is zero, the trials terminate and T becomes t. Otherwise, the weight vector w^{t+1} for the next trial is generated by (1) multiplying the weights of instances that C^t classifies correctly by the factor $\beta^t = \epsilon^t/(1 - \epsilon^t)$ and (2) renormalizing so that $\sum_x w_x^{t+1}$ is 1. The boosted classifier C^* is obtained by summing the votes of the classifiers $C^1, C^2, \ldots, C^T$, where the vote for classifier C^t is worth $\log(1/\beta^t)$ units.

Freund and Schapire [22] introduced a variant of boosting algorithm, called AdaBoost, which theoretically can significantly reduce the error of any learning algorithm if its performance is a little better than random guessing. A version of AdaBoost, i.e., AdaBoost.R, returns the weighted median as the final regression [17].

2.2.7 Gröbner Bases

This section explains the deductions used to handle geometric concepts (geometric reasoning), as well as algebraic methods relating to these. Here, "geometric concepts" refer to geometrical information, such as lines and points in Euclidean space (or another appropriate space with a distance function), as well as topological information, such as intersections in the case of straight lines or constraints in the case of surfaces.

The use of algebraic methods for theorem proofs and deductions has been known to mathematics for a long time, at latest since Alfred Tarski in the 1930s. Tarski showed that Euclidean geometry is decidable (i.e., there exists an algorithm that can decide the truth of a proposition within a finite number of steps). Unfortunately, Tarski's method is complex, required a large number of calculations, and was not actually used. However, his method has been revived in the form of the method by Collins et al. discussed below, who improved the efficiency of the method [35].

Representative examples of methods for algebraic inferences and proofs include the following.

- Wu's theorem,
- Gröbner bases,
- Quantifier Elimination(QE).

For each of these different methods, the domain targeted for reasoning varies (in a mathematical sense) as follows, and the scopes for which the completeness of proofs is guaranteed do not align perfectly.

- (Wu) Algebraic varieties formed by the roots of a polynomial,
- (Gröbner) Ideals generated by polynomials,
- (QE) Manifolds formed by the real roots of a polynomial.

Accordingly, these methods have strengths and weaknesses depending on the scope of application.

QE (Quantifier elimination) is to simplify a statement in mathematical logics by means of eliminating quantifiers. CAD (Cylindrical Algebraic Decomposition) is a QE method for the elementary theory of real closed fields (refer to [11] for details).

Wu's method has been successfully applied to geometric reasoning tasks, such as robotics, computer graphics, and computer vision (see [32] for details). For instance, it can solve the following two-circle problem (Fig. 2.7).

> Find the locus of the mid-point P of all segments QT's, where the end points Q and T lie on O_1 and O_2, respectively (O_1 and O_2 are two circles outside each other and the radii are r_1 and r_2, respectively).[2]

Recently, a field known as "algebraic biology" has been proposed, and there is vigorous research attempting to analyze the mechanisms of various biological phenomena via algebraic approaches. For example, Laubenbacher and Stigler [42] have attempted to use Gröbner bases to make deductions about genetic networks (specifically, reducing causal relationships expressed by using Boolean algebra; see Sect. 5.1 for details of GRN).

---

[2]The point P is in the range which is outside a circle (the center is a mid-point of centers of circle O_1 and circle O_2, and the radius is $\frac{|r_1-r_2|}{2}$), and which is inside a circle (the center is the same, and the radius is $\frac{(r_1+r_2)}{2}$).

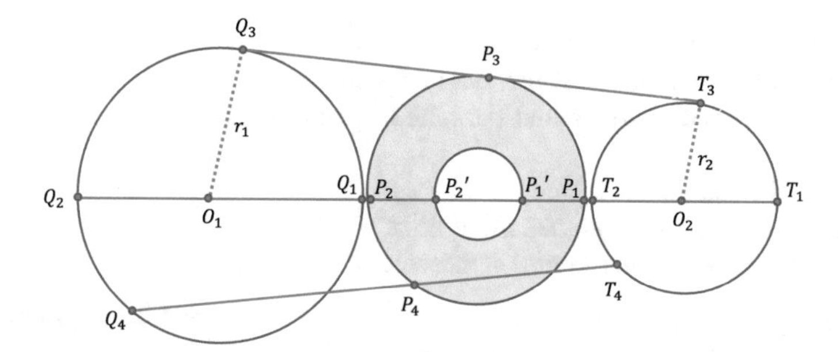

Fig. 2.7 Two-circle problem

Gröbner bases have been employed to analyze semantic building blocks in GP [67]. In the rest of this section, we explain ideas and Gröbner bases. These are relatively easy to implement and used in the extended version of GP (see Sect. 4.2).

Given a set of polynomials $F = \{f_1, \ldots, f_n\}$, the ideal generated from F

$$Ideal(F) = \{f \mid f = a_1 f_1 + \cdots + a_n f_n, \text{ where } a_i \text{ is an arbitrary polynomial}\} \tag{2.57}$$

can be represented as either $(f_1, \ldots, f_n)$ or (F). Thus, evaluating whether a certain polynomial g belongs to $(f_1, \ldots, f_n)$ corresponds to proving the truth of proposition g under assumptions $\{f_1, \ldots, f_n\}$. Gröbner bases are critical for this evaluation. We shall explain the method for proofs using Gröbner bases.

The following explanation begins with discussing several important items, and then proceeds along the lines of Buchberger [5–10].

Define a total ordering ($<_T$) for each simple term (i.e., monomial) in a polynomial. Any ordering that meets the following two conditions is acceptable.

(T1) $1 <_T t$ for all $t \neq 1$
(T2) If $s <_T t$ then $su <_T tu$ for all u

The following are commonly used examples of bivariate total orders $<_T$ (where $x <_T y$).

- total degree ordering (written $<_{TD}$)

$$1 <_{TD} x <_{TD} y <_{TD} x^2 <_{TD} xy <_{TD} y^2 <_{TD} x^3 <_{TD} x^2y <_{TD} xy^2 <_{TD} y^3 <_{TD} \cdots \tag{2.58}$$

- purely lexicographical ordering (written $<_{PL}$)

$$1 <_{PL} x <_{PL} x^2 <_{PL} x^3 < \cdots y <_{PL} xy <_{PL} x^2y < \cdots <_{PL} y^2 <_{PL} xy^2 < \cdots \tag{2.59}$$

Next, we define the following functions. In this definition, f, g, and so on are polynomials; F and G are sets of polynomials; and x and y are variables.

- $Coeff(g, t)$: the coefficient of term t in polynomial g,
- $LPP(f)$: the leading power product in f with respect to $<_T$,
- $LM(f)$: the leading monomial (i.e., the largest monomial) in f with respect to $<_T$,
- $LC(f)$: the leading coefficient of $LPP(f)$,
- $LCM(f, g)$: the least common multiple of f and g,
- $GCD(f, g)$: the greatest common measure of f and g.

We then define the following reduction rule, which acts to rewrite polynomials by replacing the highest-order term with a lower-order term.

Definition 2.1 (*Reduction*) g is reduced to h $mod(F)$: $g \rightarrow_F h \Leftrightarrow \exists f \in F, \exists b, \exists u$ s.t. $g \rightarrow_{f,b,u}$ and $h = g - buf$, where $g \rightarrow_{f,b,u}$ (g is reducible by f, b and u) is equivalent to the following two conditions being fulfilled:

$$Coeff(g, uLPP(f)) \neq 0, \tag{2.60}$$

$$b = Coeff(g, uLPP(f))/LC(f). \tag{2.61}$$

In other words, $g \rightarrow_F h$ indicates that a monomial in g is canceled out by the appropriate product buf for a certain polynomial f in F.

Consider, for example, the case where $F := \{f_1, f_2, f_3\}$.

$$F := \{f_1, f_2, f_3\} \tag{2.62}$$

$$f_1 := 3x^2y + 2xy + y + 9x^2 + 5x - 3 \tag{2.63}$$

$$f_2 := 2x^3y - xy - y + 6x^3 - 2x^2 - 3x + 3 \tag{2.64}$$

$$f_3 := x^3y + x^2y + 3x^3 + 2x^2 \tag{2.65}$$

Here, f_1, f_2, and f_3 are ordered by $<_{PL}$, so that

$$\begin{aligned} LPP(f_1) &= x^2y \quad & LC(f_1) &= 3 \\ LPP(f_2) &= x^3y \quad & LC(f_2) &= 2 \\ LPP(f_3) &= x^3y \quad & LC(f_3) &= 1 \end{aligned}$$

Now consider the reduction $mod(F)$ of the following polynomial g.

$$g = 5y^2 + 2x^2y + \frac{5}{2}xy + \frac{3}{2}y + 8x^2 + \frac{3}{2}x - \frac{9}{2} \tag{2.66}$$

Thus, we have $g \rightarrow_F h$ for the following h.

$$h = 5y^2 + \frac{7}{6}xy + \frac{5}{6}y + 2x^2 - \frac{11}{6}x - \frac{5}{2} \tag{2.67}$$

This is because the following conditions hold.

$$
\begin{aligned}
g \rightarrow_{f,b,u} \text{ where } f := f_1, b := \frac{2}{3}, u := 1 \\
Coeff(g, 1x^2y) = 2 \neq 0 \\
b = Coeff(g, 1 \cdot x^2y)/LC(f_1) \\
h = g - \left(\frac{2}{3}\right) \cdot 1 \cdot f_1
\end{aligned}
$$

Definition 2.2 (*Normal*) h is normal $mod(F) \Leftrightarrow$ there is no h' such that $h \rightarrow_F h'$

Definition 2.3 (*Normal form*) h is the normal form of $g\, mod(F) \Leftrightarrow$ h is normal and the following reduction series exists: $g = k_0 \rightarrow_F k_1 \rightarrow_F k_2 \rightarrow_F \cdots \rightarrow_F k_m = h$.

In the previous example, h is normal $mod(F)$.

Given a set of polynomials F and a polynomial g, an algorithm that finds the normal form of g is referred to as a normalization algorithm and denoted $S(F, g)$. The following is an example of a normalization algorithm S.

Algorithm 2.2 Normalization algorithm: $h := NormalForm(F, g)$

Let $h := g$.
while there exists an $f \in F$, b and a u such that $h \rightarrow_{f,b,u}$ **do**
  select f, b, u such that $uLPP(f)$ is maximal with respect to $<_T$.
  let $h := h - buf$.
end while

Definition 2.4 (*Gröbner base*) F is a Gröbner base $\Leftrightarrow$ For all g, h_1, h_2, if h_1 and h_2 are both normal forms of $g\, mod(F)$ then $h_1 = h_2$.

The method for calculating Gröbner bases was invented in 1965 by Buchberger, a student of Gröbner. This method is called as the Buchberger–Möller (BM) algorithm [49]. Furthermore, it is known that the computation of the Gröbner base or the border basis from points can be executed in polynomial time [10, 45]. Recently, the approximate Buchberger–Möller (ABM) algorithm [45] was proposed as a stable numerical method for computing the border basis [50].

The key points of BM algorithm are discussed below. Before that, the following theorems are important as preparation.

Theorem 2.1 Gröbner properties and theorems
If S is an arbitrary normalization algorithm, then the following propositions are equivalent.

(GB1) F is a Gröbner base

(GB2) $f - g \in Ideal(F) \Leftrightarrow S(F, f) = S(F, g)$

(GB3) $\rightarrow_F$ has the Church–Rosser property

The Church–Rosser property ensures that if a polynomial can be reduced by more than one method then the final result will be uniquely determined, independent of the method used.

(GB3) implies similarities between the use of Gröbner bases and term-rewriting systems or the Knuth–Bendix method [47]. In fact, S-polynomials (defined below) can be thought of as being the same as the critical pairs in the completion algorithm [7].

Definition 2.5 (*S-polynomial*) The S-polynomial for polynomials f_1 and f_2 is defined as follows: $SP(f_1, f_2) := u_1 f_1 - (c_1/c_2)u_2 f_2$, where $c_i = LC(f_i)$, $s_i = LPP(f_i)$, and $u_i = LCM(s_1, s_2)/s_i$.

For polynomials f_1 and f_2 in Eq. (2.62), for example,

$$SP(f_1, f_2) = 2x^2y + \frac{5}{2}xy + \frac{3}{2}y + 8x^2 + \frac{3}{2}x - \frac{9}{2}. \tag{2.68}$$

Theorem 2.2 Algorithmic properties of Gröbner bases
If S is an arbitrary normalization algorithm, then the following propositions are equivalent.

(GB1) F is a Gröbner base

(GB4) For all $f_1, f_2 \in F$, $S(F, SP(f_1, f_2)) = 0$

Given a set of polynomials F, the algorithm for finding G such that $Ideal(F) = Ideal(G)$ and G is a Gröbner base is as follows.

Let us try to find the Gröbner base for Eq. (2.62). First, select $\{f_1, f_2\}$ and then

$$SP(f_1, f_2) = 2x^2y + \frac{5}{2}xy + \frac{3}{2}y + 8x^2 + \frac{3}{2}x - \frac{9}{2}. \tag{2.69}$$

Reducing this polynomial yields

Algorithm 2.3 Buchberger–Möller(BM) algorithm for finding a Gröbner base

Let $G := F$.
Let $B = \{\{f_1, f_2\} \mid f_1, f_2 \in G, f_1 \neq f_2\}$.
while B is not the empty set **do**
  $\{f_1, f_2\}$:= a pair of polynomials in B.
  $B := B - \{\{f_1, f_2\}\}$.
  $h := SP(f_1, f_2)$.
  $h' := NormalForm(G, h)$.
  if $h' \neq 0$ **then**
    $B := B \cup \{\{g, h'\} \mid g \in G\}, G := G \cup \{h'\}$.
  end if
end while

$$\frac{7}{6}xy + \frac{5}{6}y + 2x^2 - \frac{11}{6}x - \frac{5}{2}. \tag{2.70}$$

Multiply this polynomial by $\frac{6}{7}$ so that LC becomes 1 to produce the following polynomial f_4, which is then added to G.

$$f_4 = xy + \frac{5}{7}y + \frac{12}{7}x^2 - \frac{11}{7}x - \frac{15}{7} \tag{2.71}$$

Then, selecting $\{f_1, f_4\}$ gives

$$SP(f_1, f_4) = 1 \cdot f_1 - \frac{3}{1} \cdot x \cdot f_4 \tag{2.72}$$

$$= -\frac{1}{7}xy + y - \frac{36}{7}x^3 + \frac{96}{7}x^2 + \frac{80}{7}x - 3. \tag{2.73}$$

Reducing this polynomial by subtracting $-(1/7)f_4$ and then adjusting so that LC becomes 1 we get

$$f_5 := y - \frac{14}{3}x^3 + \frac{38}{3}x^2 + \frac{61}{6}x - 3. \tag{2.74}$$

Further subtracting $(5/7)f_5$ from $SP(f_4, f_5) = 1f_4 - (1/1)xf_5$ and adjusting so that LC becomes 1 gives

$$f_6 := x^4 - 2x^3 - \frac{15}{4}x^2 - \frac{5}{4}x. \tag{2.75}$$

Finally, from $SP(f_1, f_3) = xf_1 - (3/1)1f_3$ we obtain

$$f_7 := x^3 - \frac{5}{2}x^2 - \frac{5}{2}x. \tag{2.76}$$

SP reduction of the remaining pairs yields 0 in all cases, and so the derivation of the basis is complete. For example,

$$SP(f_6, f_7) = \frac{1}{2}x^3 - \frac{5}{4}x^2 - \frac{5}{4}x, \tag{2.77}$$

but by subtracting $(1/2) f_7$ this can be reduced to 0. Thus, the Gröbner base obtained is $G := \{f_1, f_2, \ldots, f_7\}$.

Definition 2.6 (*Reduced Gröbner base*) F is a (reduced)Gröbner base $\Leftrightarrow$ F is a Gröbner base and $LC(F) = 1$ for all $f \in F$ and f is normal $mod(F - \{f\})$

Given a Gröbner base, it is possible to obtain a reduced Gröbner base by first removing polynomials that can be reduced to 0 by other polynomials and then normalizing the remaining polynomials so that LC becomes 1.

In the equation above, $G = \{f_1, f_2, \ldots, f_7\}$ is not a reduced Gröbner base. This is because f_1 can be reduced to 0 using $mod(\{f_2, \ldots, f_7\})$. Furthermore, f_1, f_2, f_3, f_4 and f_6 are reduced to 0 and f_5 is reduced to the following polynomial:

$$f_5' := y + x^2 - \frac{3}{2}x - 3. \tag{2.78}$$

Thus, the reduced Gröbner base becomes

$$G' := \{f_5', f_7\} \tag{2.79}$$

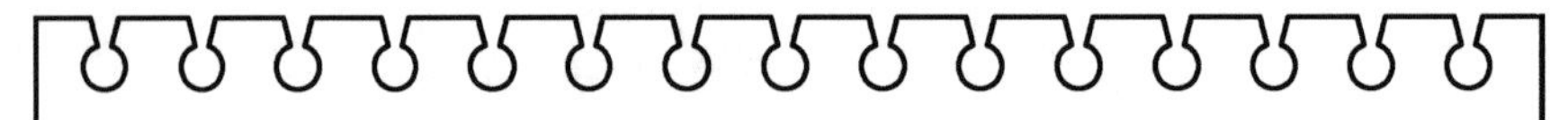

Theorem 2.3 Uniqueness of reduced Gröbner bases
Suppose $Ideal(F) = Ideal(F')$ and both F and F' are reduced Gröbner bases. In this case, F and F' agree as sets.

Let GB be a function that returns G meeting the following conditions for the set of polynomials F.

- $G := GB(F)$,
- $Ideal(F) = Ideal(G)$,
- G is a reduced Gröbner base.

Then, the following key theorem holds true for GB.

Theorem 2.4 Key theorem for reduced Gröbner bases [8, p.194]
GB fulfills the following proposition for all F and G.

(SGB1) $Ideal(F) = Ideal(GB(F))$.

(SGB2) If $Ideal(F) = Ideal(G)$ then $GB(F) = GB(G)$.

(SGB3) $GB(F)$ is a reduced Gröbner base.

Gröbner bases play an important role in calculating ideal bases. The following section gives some example applications of Gröbner bases.

2.2.7.1 Evaluating Ideal Matching and Membership

Consider the following kinds of problems.

1. Given F, f, and g, evaluate whether $f - g \in Ideal(F)$.
2. Given F and f, evaluate whether $f \in Ideal(F)$.
3. Given $F_1$and F_2,evaluate whether $Ideal(F_1) \subseteq Ideal(F_2)$.

These questions can be answered using Gröbner bases as follows.

1. Suppose $G := GB(F)$ then

$$f - g \in Ideal(F) \Leftrightarrow S(G, f) = S(G, g). \tag{2.80}$$

2. Suppose $G := GB(F)$ then

$$f \in Ideal(F) \Leftrightarrow S(G, f) = 0. \tag{2.81}$$

3. Suppose $G_2 := GB(F_2)$ then

$$Ideal(F_1) \subseteq Ideal(F_2) \Leftrightarrow \forall f \in F_1 : S(G_2, f) = 0. \tag{2.82}$$

2.2.7.2 Solvability of Algebraic Expressions

Let us evaluate whether a given F is solvable. Being solvable means that there is some algebraic extension K such that all f in F satisfy $f(a_1, \ldots, a_n) = 0$ for the elements a_i of K, where F is a set of n variable polynomials over $x_1, x_2, \ldots, x_n$. Suppose that $G := GB(F)$. Then, we just need to use the following property.

$$F : \text{unsolvable} \Leftrightarrow 1 \in G \tag{2.83}$$

For example, F in Eq. (2.62) is solvable. This is because $GB(F)$, found earlier (Eq. (2.79)), did not contain 1.

In contrast, the following F is not solvable.

$$F := \{x^2y - x^2, x^3 - x^2 + y, xy^2 - xy + 2\} \tag{2.84}$$

To investigate this, we first adopt $<_{TD}$ as the total order to find the Gröbner base.

$$SP(x^2y - x^2, x^3 - x^2 + y) = x^2y - x^3 - y^2 \rightarrow_F -x^3 - y^2 + x^2 \rightarrow_F -y^2 + y \tag{2.85}$$

Thus, we add $y^2 - y$ to the base.

$$SP(xy^2 - xy + 2, y^2 - y) = 2 \tag{2.86}$$

This cannot be reduced any further, and so 1 is included in the base. This means that this F is not solvable.

Next, we want to evaluate whether a given F has a finite number of solutions, rather than an infinite number of them, where F is a set of n variable polynomials over $x_1, x_2, \ldots, x_n$. Suppose then that $G := GB(F)$. Then, we just need to use the following property.

> F has a finite number of solutions. $\Leftrightarrow$
> For all i $(1 \leq i \leq n)$ there is a polynomial in G such that the LPP has the form $x_i^{k_i}$.

For example, the F in Eq. (2.62) has a finite number of solutions. This is because the LPPs for the polynomials in $GB(F)$ that we found above (i.e., Eq. (2.79)) are x^3 and y. In contrast, the following F has an infinite number of solutions.

$$F := \{x^2y - y^2 - x^2 + y, x^2 - y\} \tag{2.87}$$

In fact, this has an infinite number of solutions because F is already a Gröbner base but there is no polynomial with its LPP in the form y^j.

Next, given a solvable F with a finite number of solutions, let us try to find all solutions of F. Suppose that $G := GB(F)$, adopting $<_{PL}$ as the total ordering. It then follows that G includes polynomials of $K[x_1]$ only. For this, we execute the following elimination algorithm.

For example, consider F from Eq. (2.62), which was

$$GB(F) := \{f := x^3 - 5/2x^2 - 5/2x, g := y + x^2 - 3/2x - 3\}. \tag{2.88}$$

The equation $f = 0$ contains only x as a variable, and so the solutions are 0, $(5 + \sqrt{65})/4$ and $(5 - \sqrt{65})/4$. Substituting these x values into $g = 0$ to find y, we obtain three solutions.

Algorithm 2.4 Elimination algorithm

$p :=$ a polynomial in $G \cap K[x_1]$. ▷ Returns X_n including all solutions.
$X_1 := \{(a) \mid p(a) = 0\}$.
for $i = 1$ to $n - 1$ **do**
  $X_{i+1} := \phi$.
  for all $(a_1, \ldots, a_i)$ in X_i **do**
    $H := \{g(a_1, \ldots, a_i, x_{i+1}) \mid g \in G \cap K[x_1, \ldots, x_{i+1}] - K[x_1, \ldots, x_i]\}$.
    $p :=$GCD for the polynomials in H.
    $X_{i+1} := X_{i+1} \cup \{(a_1, \ldots, a_i, a) \mid p(a) = 0\}$.
  end for
end for

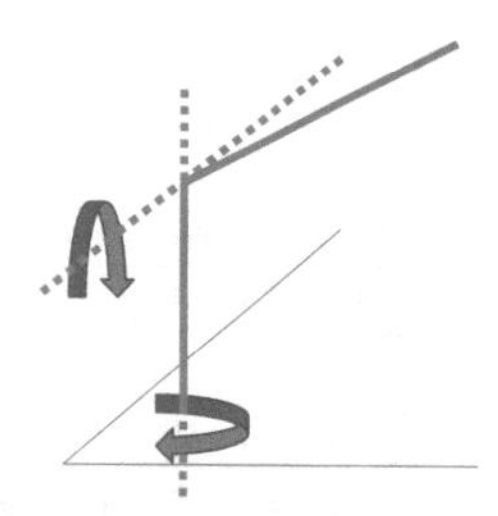

Fig. 2.8 An example of inverse robotic kinematics

2.2.7.3 Examples of Applications to Other Fields

Gröbner bases have been applied to other fields. Some examples follow.

1. Solutions for the mechanical properties of robots,
2. Calculating intersections of superellipsoids,
3. Eliminating parameters and the converse,
4. Detecting singularities,
5. Proving geometrical theorems,
6. Analyzing prime ideals.

Figure 2.8 shows a practical application to inverse robotic kinematics [63]. This problem involves calculating the constraints for distances and angles in a joint with the aim of guiding the end-effector to a given position and orientation. As an example, consider a robot with a rotating joint having two degrees of freedom, as shown in the figure. Here, we introduce variables as listed in Table 2.1. The constraints given by the figure are then as listed in Table 2.2.

These variables can be categorized as

1. Geometrical variables l_1, l_2
2. Position variables $p_x, p_y, p_z, sf, cf, st, ct$
3. Joint variables s_1, c_1, s_2, c_2

Most problems in inverse robotic kinematics can be described in the following manner.

1. The geometrical variables are given.

Table 2.1 Robot control variables

l_1, l_2	Lengths of the two robot arms
p_x, p_y, p_z	x, y, z coordinates of the end-effector
ϕ, θ, ψ	Eulerian angles in the direction of the end-effector
δ_1, δ_2	Angles of rotation for the rotating joint
s_1, c_1	sin and cos, respectively, of δ_1
s_2, c_2	sin and cos, respectively, of δ_2
sf, cf	sin and cos, respectively, of ϕ
st, ct	sin and cos, respectively, of θ
sp, cp	sin and cos, respectively, of ψ

Table 2.2 Constraints for the robot

$c_1 \cdot c_2 - cf \cdot ct \cdot cp + sf \cdot sp = 0$	$s_2 + st \cdot cp = 0$
$s_1 \cdot c_2 - sf \cdot ct \cdot cp - cf \cdot sp = 0$	$c_2 - st \cdot sp = 0$
$-c_1 \cdot s_2 - cf \cdot ct \cdot sp + sf \cdot cp = 0$	$s_1 - cf \cdot st = 0$
$-s_1 \cdot s_2 + sf \cdot ct \cdot sp - cf \cdot cp = 0$	$ct = 0$
$l_2 \cdot c_1 \cdot c_2 - px = 0$	$l_2 \cdot s_1 \cdot c_2 - py = 0$
$l_2 \cdot s_2 + l_1 - pz = 0$	$c_1^2 + s_1^2 - 1 = 0$
$c_2^2 + s_2^2 - 1 = 0$	$cf^2 + sf^2 - 1 = 0$
$ct^2 + st^2 - 1 = 0$	$cp^2 + sp^2 - 1 = 0$

2. Several of the independent position variables are also given.
3. Under these conditions, for the remaining position variables and joint variables, find the dependency relations with the other geometrical variables.

Traditionally, these problems have been addressed using computational methods such as numerical solutions. Buchberger demonstrated a method for generalizing these problems using Gröbner bases.

Now consider the case where the geometrical variables are l_1 and l_2, and the position variables are px and pz. For this case, we find the Gröbner base on $Q(l_1, l_2, p_x, p_z)[c_1, \ldots, sp]$, where Q is rational number field.

For the ordering, use

$$c_1 < c_2 < s_1 < s_2 < p_y < cf < ct < cp < sf < st < sp \tag{2.89}$$

The results are shown in Table 2.3.

The Gröbner base obtained in this manner has the following advantages.

1. l_1, l_2, p_x, p_z remain as variables, and so the solution is more general than that for numerical computation methods.

Table 2.3 Gröbner base obtained

$c_1^2 + \frac{p_x^2}{p_z^2 - 2 \cdot l_1 \cdot p_z - l_2^2 + l_1^2} = 0$	$c_2 + \frac{p_z^2 - 2 \cdot l_1 \cdot p_z - l_2^2 + l_1^2}{l_2} \cdot p_x \cdot c_1 = 0$
$s_1^2 - \frac{p_z^2 - 2 \cdot l_1 \cdot p_z + p_x^2 - l_2^2 + l_1^2}{p_z^2 - 2 \cdot l_1 \cdot p_z - l_2^2 + l_1^2} = 0$	$s_2 - \frac{p_z - l_1}{l_2} = 0$
$p_y + \frac{p_z^2 - 2 \cdot l_1 \cdot p_z + p_x^2 - l_2^2 + l_1^2}{p_x} \cdot c_1 \cdot s_1 = 0$	$c_t = 0$
$cf^2 - \frac{p_z^2 - 2 \cdot l_1 \cdot p_z + p_x^2 - l_2^2 + l_1^2}{p_z^2 - 2 \cdot l_1 \cdot p_z - l_2^2 + l_1^2} = 0$	
$cp + \frac{p_z^3 - 3 \cdot l_1 \cdot p_z^2 - l_2^2 \cdot p_z + 3 \cdot l_1^2 \cdot p_z + l_1 \cdot l_2^2 - l_1^3}{l_2 \cdot p_z^2 - 2 \cdot l_1 \cdot l_2 \cdot p_z + l_2 \cdot p_x^2 - l_2^3 + l_1^2 \cdot l_2} \cdot s_1 \cdot cf = 0$	
$sf + \frac{p_z^2 - 2 \cdot l_1 \cdot p_z - l_2^2 + l_1^2}{p_z^2 - 2 \cdot l_1 \cdot p_z + p_x^2 - l_2^2 + l_1^2} \cdot c_1 \cdot s_1 \cdot cf = 0$	
$st + \frac{p_z^2 - 2 \cdot l_1 \cdot p_z - l_2^2 + l_1^2}{p_z^2 - 2 \cdot l_1 \cdot p_z + p_x^2 - l_2^2 + l_1^2} \cdot s_1 \cdot cf = 0$	
$sp + \frac{p_z^4 - 4 \cdot l_1 \cdot p_z^3 - 2 \cdot l_2^2 \cdot p_z^2 + 6 \cdot l_1^2 \cdot p_z^2 + 4 \cdot l_1 \cdot l_2^2 \cdot p_z - 4 \cdot l_1^3 \cdot p_z + l_2^4 - 2 \cdot l_1^2 \cdot l_2^2 + l_1^4}{l_2 \cdot p_x \cdot p_z^2 - 2 \cdot l_1 \cdot l_2 \cdot p_x \cdot p_z + l_2 \cdot p_x^3 - l_2^3 \cdot p_x + l_1^2 \cdot l_2 \cdot p_z} \cdot c_1 \cdot s_1 \cdot cf = 0$	

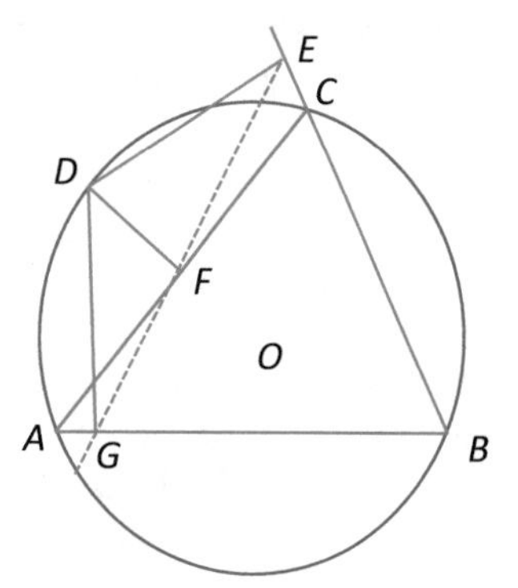

Fig. 2.9 Simson's theorem

2. The base obtained has been triangularized. In other words, the basis is in a form in which the first expression depends only on c_1, the second expression depends on c_1 and c_2, the third expression depends on c_1, c_2, and s_1, and so on. This depends on the general property discussed on p. 56 (ideal elimination, the fact that G includes polynomials of $K[x_1]$ only).
3. The degree of the base does not become high. This property does not always hold, but it has been demonstrated that the non-leading expressions (the second and subsequent expressions) are generally low degree.
4. All solutions can be obtained by numerical substitution. Moreover, no extraneous solutions are derived.

Figure 2.9 shows the proof of a geometrical theorem using a Gröbner base [14]. This is the proof of Simson's theorem.

Table 2.4 Expressions for the assumptions and conclusion

$h_1 = 2u_1x_1 - u_1^2 = 0$	OA = OB
$h_2 = 2u_3x_2 + 2u_2x_1 - u_3^2 - u_2^2 = 0$	OA = OC
$h_3 = -x_3^2 + 2x_2x_3 + 2u_4x_1 - u_4^2 = 0$	D is on circle O
$h_4 = u_3x_5 + (u_2 - u_1)x_4 - u_3x_3 + (-u_2 + u_1)u_4 = 0$	DE ⊥ BC
$h_5 = (-u_2 + u_1)x_5 + u_3x_4 - u_1u_3 = 0$	E, B, and C are collinear
$h_6 = u_3x_7 + u_2x_6 - u_3x_3 - u_2u_4 = 0$	DF ⊥ AC
$h_7 = -u_2x_7 + u_3x_6 = 0$	F, A, and C are collinear
$g = (-x_4 + u_4)x_7 + x_5x_6 - u_4x_5 = 0$	

Simson's theorem
Suppose D is a point on the circumcircle around △ABC, and E, F, and G are the perpendicular feet formed with line segments from D to the (extended) sides BC, CA, and AB. Then, E, F, and G are collinear, all lying on a line known as the Simson line.

Here we set up a coordinate system as follows.

$$A = (0, 0) \quad B = (u_l, 0) \quad C = (u_2, u_3) \quad O = (x_l, x_2)$$
$$D = (u_4, x_3) \quad E = (x_4, x_5) \quad F = (x_6, x_7) \quad G = (u_4, 0)$$

Having this, we can write expressions for the assumptions as $h_1, \ldots, h_7$ and an expression for the conclusion as g, as shown in Table 2.4. When we find a Gröbner base for $h_1, \ldots, h_7$ over $Q(u_1, \ldots, u_4)$ (where Q is rational number field), the result is as shown in Table 2.5. Here, g can be rewritten to 0 via the Gröbner base above, thus proving the proposition.

2.2.8 *Affinity Propagation and Clustering Techniques*

Affinity propagation (AP) uses a process called "message passing" to recursively calculate values (messages) between data points until the process converges [23]. A message is an evaluation value sent from one data point to another, with higher values indicating better evaluations. AP uses two types of messages, "responsibility" and "availability," derived on the basis of a similarity function that indicates the degree of similarity between pairs of data points. Message passing refers to the process of exchanging these two types of messages between all data points. Then, based on the values of the messages that result from convergence during message passing, the algorithm determines which data points best represent the clusters as exemplars, as

Table 2.5 Gröbner base obtained

$x_7 - (u_3^2 x_3 + u_2 u_3 u_4)/(u_3^2 + u_2^2)$
$x_6 - (u_2 u_3 x_3 + u_2^2 u_4)/(u_3^2 + u_2^2)$
$x_5 - (u_3^2 x_3 + ((u_2 - u_1)u_3)u_4 + (-u_1 u_2 + u_1^2)u_3)/(u_3^2 + u_2^2 - 2u_1 u_2 + u_1^2)$
$x_4 - (((u_2 - u_1)u_3)x_3 + (u_2^2 - 2u_1 u_2 + u_1^2)u_4 + u_1 u_3^2)/(u_3^2 + u_2^2 - 2u_1 u_2 + u_1^2)$
$x_3^2 - ((u_3^2 + u_2^2 - u_1 u_2)x_3 - u_3 u_4^2 + u_1 u_3 u_4)/u_3$
$x_2 - (u_3^2 + u_2^2 - u_1 u_2)/(2u_3)$
$x_1 - u_1/2$

well as which exemplar clusters should contain each of the remaining data points. This means that with AP the number of clusters and the form of the clusters are determined automatically because all data points have the possibility of becoming an exemplar and messages are passed in such a way that the similarity between exemplars and the other members of their clusters becomes as high as possible.

Similarity characterizes the degree of similarity between pairs of data points, expressed as $s(i, j)(i, j = 1, 2, \ldots, N)$, with larger values indicating a higher degree of similarity. Any index can be used for similarity so long as it meets the condition that larger values indicate greater similarity, so a range of functions can be used, such as Euclidean distance, negative or reciprocal of the squared error, or the Jaccard coefficient.[3] The $s(i, i)$ values of similarity are referred to as the preferences, and the number of clusters varies according to the magnitude of the diagonal. Usually, the median similarity value of each element is used for its preference.

Responsibility is a message that is sent from cluster member i to a data point k that is a candidate exemplar. The value of this message indicates how appropriate k is as an exemplar for i. The responsibility message is given as follows (see Fig. 2.10):

$$r(i, k) \leftarrow s(i, k) - \max_{k' s.t. k' \neq k} (a(i, k') + s(i, k')). \tag{2.90}$$

The larger the availability value from a data point k' (a point other than k) to data point i, the smaller the value of Eq. (2.90). In other words, if a data point i receives a high evaluation from a point other than k, the value of the responsibility from i to k becomes smaller.

Availability is a message sent from data point k (an exemplar candidate) to data point i, which is a candidate member of the same cluster as k. This message indicates

[3] A degree measure of the similarity between two documents A and B, i.e., $\frac{|A \cap B|}{|A \cup B|}$.

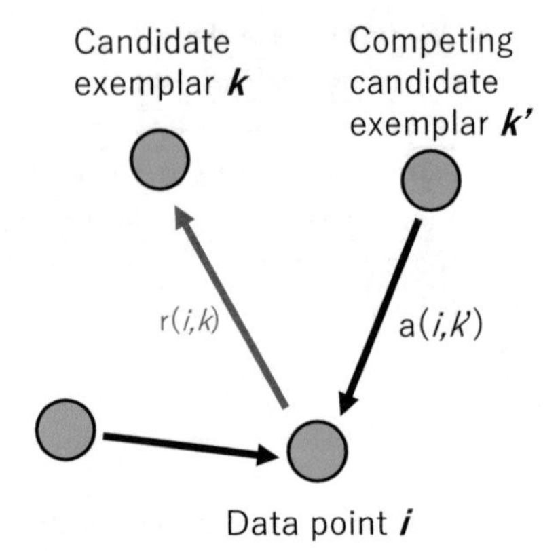

Fig. 2.10 Sending responsibilities

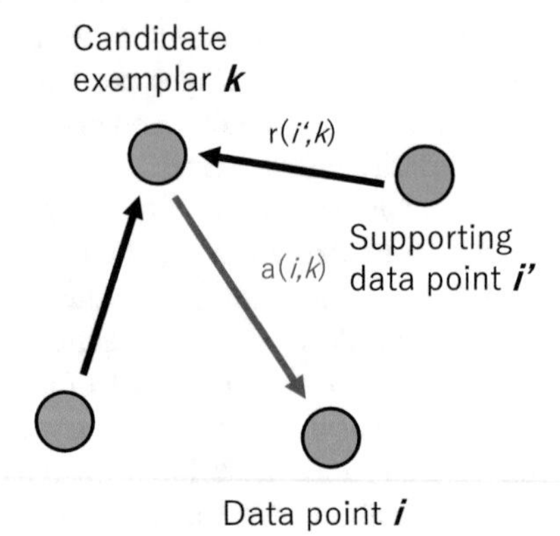

Fig. 2.11 Sending responsibilities

the appropriateness of data point i being grouped with the exemplar candidate k. The availability message is given as follows (see Fig. 2.11):

$$a(i,k) \leftarrow s(i,k) - \min\left\{0, r(k,k) + \sum_{i' s.t. i' \notin i,k} \max\{0, r(i',k)\}\right\} \tag{2.91}$$

$$a(k,k) \leftarrow \sum_{i' s.t. i' \notin i,k} \max\left\{0, r(i',k)\right\} \tag{2.92}$$

Equations (2.91) and (2.92) use the sum of the positive responsibilities from data points i' (i.e., points other than i) to data point k. This represents how high an evaluation data point k has received from other data points. In other words, the greater the popularity of k as a node, the greater the value of the availability from k to i.

Algorithm 2.5 Calculation procedure for affinity propagation

Initialize R(responsibility),A(availability) with 0
while termination criteria are not met **do**
  Update R by Eq. (2.90)
  Update A by Eqs. (2.91) and (2.92)
  Find exemplar k, where k satisfies Eq. (2.93)
end while ▷ End of clustering

AP finds exemplars by recursively calculating responsibilities and availabilities. Data points becoming exemplars are those that satisfy the following relationship:

$$r(k,k) + a(k,k) > 0. \tag{2.93}$$

The recursion terminates either when there is no change in the exemplar nodes for a certain number of iterations or when the depth of recursion reaches a certain limit. The non-exemplar data points i are assigned to the cluster whose exemplar is the data point k such as $r(i,k) + a(i,k)$ is a maximum. Algorithm 2.5 shows an overview of the calculation procedure for AP.

Figure 2.12 shows an example process of affinity propagation.[4] Different colors mean different clusters. We can observe the self-organization of clusters with iterations. Note that different numbers of clusters are gradually formed. Thus, the user need not set the number of clusters beforehand.

Compared with the previous clustering methods such as the k-means algorithm (Sect. 2.2.1), AP has several advantages as follows:

- There is no need to decide the number of clusters in advance, as this is determined automatically by the algorithm.
- There is no dependence on initial values, unlike k-means where the results vary depending on how the initial values for the centers of clusters are selected.
- The constraints on the similarity function $s(i, j)$ are loose: $s(i, j)$ does not have to be commutative and does not need to satisfy the triangle inequality.

Accordingly, AP has been used with a large number of methods, including active learning [30], key image extraction [46], image clustering [18, 34], and extracting representative images [72].

[4]The used parameters are as follows: 300 sample points, cluster centers = [[2, 2], [−2, −2], [2, −2]], and cluster std. = 0.5.

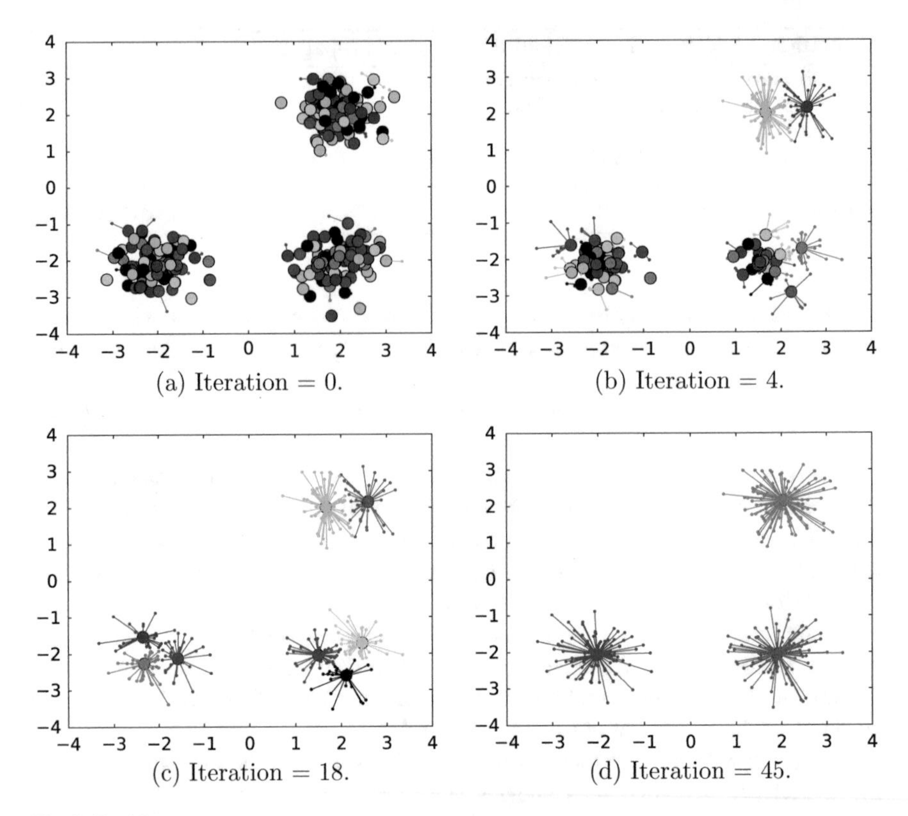

(a) Iteration = 0. (b) Iteration = 4.

(c) Iteration = 18. (d) Iteration = 45.

Fig. 2.12 AP process

2.3 Deep Learning Frameworks

2.3.1 CNN and Feature Extraction

When applied to image processing or text recognition, conventional neural networks tend to produce poor results in case of being translated, rotated, or distorted relative to the original training data. This is because these neural networks are trained using only raw data and ignore the topological features of the input data.

Convolutional neural networks (CNNs) were proposed to address this weakness. CNNs are based on the insight from neurobiology that the visual cortex includes neurons, the LGN-V1-V2-V4-IT layers (the ventral stream), that are both locationally and directionally selective. CNNs also hold promise as models of object recognition in the visual cortex. This approach involves modeling vision in terms of local receptive fields, simple cells, and complex cells. CNNs have their origins in the neocognitron approach proposed by Kunihiko Fukushima [24]. Another early model is LeNet [44].

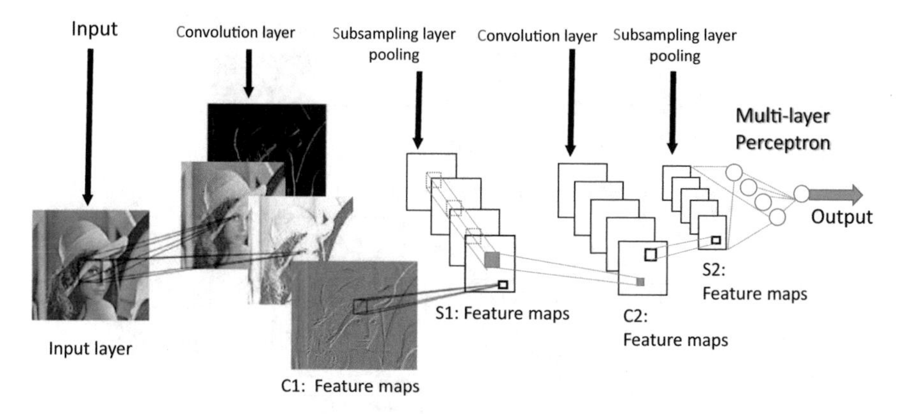

Fig. 2.13 Example of a convolutional neural network (CNN)

CNNs are a type of multilayer feed-forward neural network that can extract topological features from images, with training performed via backpropagation. The networks are designed so as to be able to recognize visual patterns directly from pixel images without hardly any pre-processing. This also allows handling patterns that can vary widely, such as handwritten text.

Figure 2.13 shows the configuration of a CNN, consisting of recursive C and S layers, as described below:

- C layers: Convolution layers to extract features.
- S layers: Pooling layers to achieve topological invariance with respect to shifts and distortions.

Each layer combines patterns from the preceding layer by integrating or smoothing them. Typically, a CNN can compress a large image into a small number of features that are robust to local deformation.

The C layers detect the same feature of an input image at different locations. This means that all of the neurons relating to a certain feature share the same weights (but with different biases). In this way, the preceding neuron responds to the same feature at different locations in the input. Then, if a certain neuron in a feature map fires, this is like matching a template (a pattern for extracting a feature). This is where convolution takes place (see Fig. 2.14). Convolution is a binary operation that calculates a sliding weighted sum of two functions, putting one function over another and summing the overlap. In two-dimensional image processing, this corresponds to multiplying the pixel values in the original image by a weighting matrix (known variously as a filter, mask, or kernel) and summing to obtain the results. This enables, for example, the original image to be softened or sharpened. Convolution is typically applied to a given pixel and its surrounding values, using a small filter matrix, often 3×3 pixels in size. This processing can be performed extremely quickly because pixels can be processed in parallel. The C layer writes the results of the convolutions.

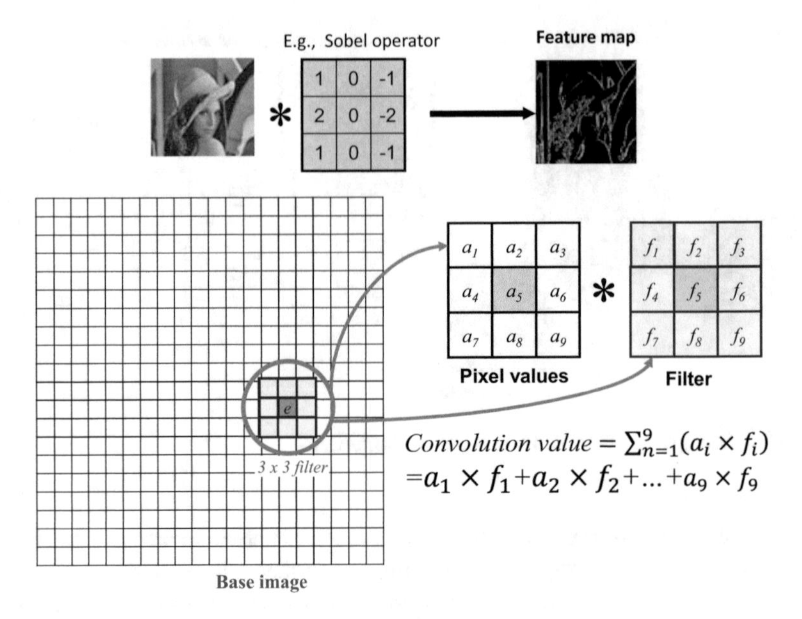

Fig. 2.14 Example of convolution

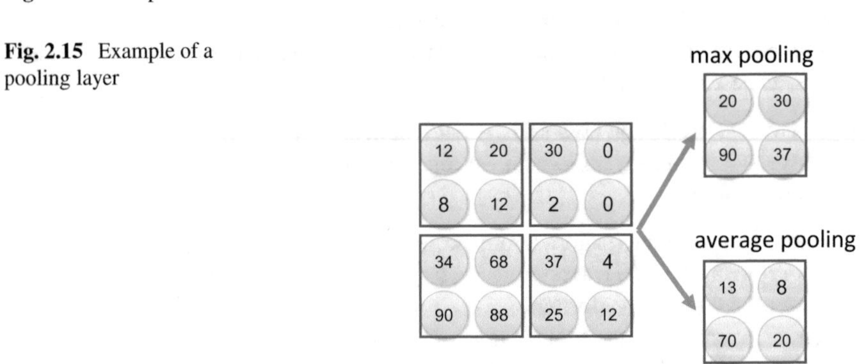

Fig. 2.15 Example of a pooling layer

This means that multiple feature maps can be generated by preparing a variety of filters.

The S layer reduces the spatial sensitivity of the feature map (Fig. 2.15). This makes it possible to achieve robustness against a certain degree of shifts and distortions. The goal is to group features with similar meanings. Weight sharing is applied at this layer as well, and this makes it possible to reduce the effects of noise. The S layer performs data compression and smoothing. In other words, this achieves robustness so that the output to the C layer is the same even when there are variations in the input to the S layer, such as small shifts. Normally, this processing finds the average or maximum value of a pattern without any overlaps. This layer of processing is known as the pooling layer because it pools the output from the S layer neurons and outputs it to the C layer.

These two layers are normally created manually rather than trained. Thereafter, supervised learning is performed using training data [44]. This means that there is a requirement for a certain degree of regularity and continuity in the input data. Real-world images have these kinds of features, and so statistical features can be expected to persist.

Let us take another look at the CNN processing shown in Fig. 2.13. The input image is convolved using three trainable filters and biases to generate four feature maps in the C1 layer. Each group of four pixels in the feature map is weighted and added to the bias values to generate four feature maps in the S1 layer via a sigmoid function. Filters are once again applied to these to generate the C2 level. S2 is generated in the same way as S1, and finally these pixel values are passed as input to a normal neural network.

Sparse coding, i.e., a numerical model of processing biological visual information, is related to the CNN mechanism. Sparse coding uses a small set of basis vectors chosen such that the given images can be expressed as linear combinations of these basis vectors. In other words, sparse coding extracts a small amount of information essential for forming a concept from the large amount of information that has been input to the neural network. As CNNs are given hidden layers of increasing depth, they express concepts with higher dimensionality, moving from the points in an image to lines, from lines to outlines, from outlines to parts, and from parts to the whole image.

Some researchers have attempted to train CNNs to extract high-level features, such as cats or human faces, from large volumes of images [43]. The network used in [43] had nine layers, consisting of three subnetworks, each with three layers and a pooling function. Ten million images were input, and the network was trained using an autoencoder [28]. As a result, the network succeeded in recognizing a wide range of objects, such as faces and cats. This recognition device also had a certain degree of robustness against variations such as rotation, scaling, or translation. Moreover, neurons have been formed that seem to respond specifically to high-level features, such as faces. The average features of many faces (a large amount of training data) are captured well. This means that this can be thought of as the internal representation of the neural network obtained by the face training.

This network is assumed to have about ten million parameters that need to be trained. In contrast, the neurons and synapses in the human visual cortex have a million times as many parameters as this. This fact should give a sense of the practicality of this model. However, it took 1000 cluster machines (with 16,000 cores) three days to train the network. This is an extraordinary amount of calculations.

In another study, a CNN categorized 1.2 million items of data in a high-resolution image library (ImageNet [40]) into a thousand different classes [40]. This network consists of 650,000 neurons and 60 million parameters, organized as five convolutional layers[5] and three globally connected layers including the softmax activation

[5]The first, second, and fifth of these layers are linked to max pooling.

function.[6] The output layer consists of 1000 nodes, one for each of the classes into which images are to be classified.

The following two methods were used to reduce overfitting, which might otherwise be a problem.

- Data augmentation: Increasing the number of training items by randomly cutting a relatively small image from the original image and then shifting or horizontally reflecting it to form the mirror image. Images where the RGB intensities have been changed slightly are also created.[7]
- Drop out: The output from nodes in hidden layers are set to 0 at random (with a probability of 0.5). Zeroed nodes do not contribute to the output of downstream nodes, and so do not participate in backpropagation. In this way, training is performed with a different architecture every time input is presented.

As a result of these improvements, the neural network achieved extremely good results in the ILSVRC2012 contest.[8] Training the network took between five and six days on two GTX 580 3GB GPUs. The performance could be improved with faster GPUs and a larger data set.

CNNs have achieved huge successes in instantaneous object recognition, unaffected by backgrounds or postures. This processing recalls similarities with object recognition in the visual cortex (the ventral stream). At the same time, there are recognition tasks that still cannot be properly handled by CNNs. Examples include recognizing where something is in space (processed by the dorsal stream rather than the ventral stream) and video recognition.

2.3.2 Generative Adversary Networks (GAN) and Generating Fooling Images

Deep learning as a generation model has yielded very interesting results in recent years. In particular, the GAN [26] proposed by Goodfellow et al. facilitates the generation of a much clearer image than the variational autoencoder, which had hitherto been known as the generation model. Thereafter, assorted variations, including the deep convolutional generative adversarial networks (DCGAN) [61] and variational autoencoder with GAN (VAEGAN) [41], have been proposed.

DCGAN is a generation model extended in a form that specializes the generative adversarial network (GAN) to image generation. In the GAN, the discriminator and

[6] The softmax activation function is defined as $f_i(x_1, \ldots, x_d) = \frac{\exp(x_i)}{\sum_j \exp(x_j)}$ $(i = 1, \ldots, d)$, where d is the unit number of the layer.

[7] Principal component analysis of the RGB values in the data set is performed and then random changes are made, centered on these axes, thereby preserving the essential features of the original natural images.

[8] ILSVRC2012 dataset is a subset of the ImageNet database [40]. It contains 1,000 object categories. The training set, validation set, and testing set contain 1.3 M, 50 K, and 150 K images, respectively. The input images are of $224 \times 224 \times 3$ pixels.

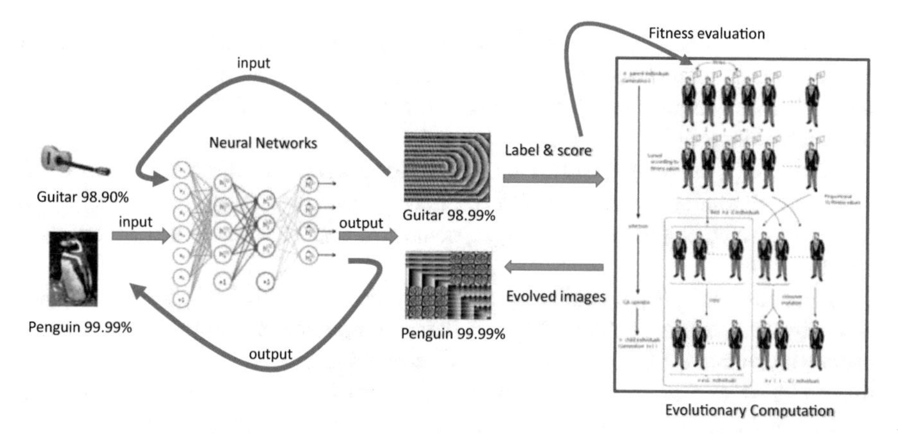

Fig. 2.16 How to fool the deep learning model

generator learning machines are used to learn the shape of the distribution itself without being given the distribution of the training data set in advance, thereby developing the generator to generate data that are indistinguishable from the learning data set. A random number z sampled from a uniform distribution is input into the generator, and x is generated with this as a seed. The discriminator, a classifier that determines whether an input is derived from the learning data set or is generated by the generator, enables the generator to learn. In the DCGAN, the discriminator uses a normal CNN and the generator uses a reverse CNN starting from z.

DNNs can recognize images with high reliability, but DNNs can be easily deceived. For example, as shown in Fig. 2.16, research conducted by Nguyen et al. [51] indicated that both the images of the actual objects (left) and patterns of the objects (right) are recognized by DNN as being either a "guitar" or "penguin" with 99.9% confidence. In other words, although the images on the right are only a meaningless pattern for human beings, DNN recognizes them as meaningful objects.

The pattern on the right is a synthesized fooling image generated by GA, and the details of the generation process are as follows.

Step1 Express images as genotypes, and carry out crossover and mutations.
Step2 The new image is identified by the DNN model, and the reliability of the recognition result is used as the fitness value.
Step3 Select individuals (i.e., images) showing a high recognition level for all classes and use them for generating of the next generation.

An ordinary GA selects only solutions with high fitness in the target class in order to search for the optimal solution. For example, when the goal is to recognize an object as a "guitar," an image recognized to be a "guitar" and having high recognition reliability is selected. Meanwhile, when generating an fooling image, an image showing high reliability in all classes including the target class is selected, rather than an image with high recognition reliability in the target class. For example, when the aim is to generate a fooling image of a guitar, an image was selected that showed high reliability in all classes, including "guitar," "violin," and "biwa" at the same time, rather than an image showing high reliability as a "guitar."

In this study, the genotype of an image based on pixel unit or CPPN (see Sect. 3.1.2) was used. Evolutionary computation resulted in the successful generation of white noise and striped pattern fooling images, demonstrating the vulnerability of the DNN model.

In our research [70], a similar process is applied to the original speech data, and an attempt is made to synthesize speech that a computer can recognize, but humans cannot understand. In this experiment, false-positive speech has been synthesized using GA. It was experimentally shown that even for sounds that are heard only as white noise by a human, computers can recognize a meaningful word with high reliability.

2.3.3 Bayesian Networks and Loopy Belief Propagation

Belief propagation (BP) [58] is an inference algorithm for tree-structured graphical models and have some variants depending on objectives. For example, max-sum is the instance of BP and calculates the highest joint probability of those models effectively. LBP (Loopy Belief Propagation) is the application of BP to graphical models with arbitrary loopy graph structure and approximately and speedily infers most probable solutions (MPSs) or marginal probabilities.

The term *loopy max-sum* is used to refer to the application of max-sum algorithm to loopy graphs. Loopy max-sum repeatedly update messages, which are locally calculated joint probabilities, and finally approximately generates MPS from the messages. In order to apply loopy max-sum to Bayesian networks, we must transform them to equivalent factor graphs, which are a type of graphical models. It is possible to transform Bayesian networks to equivalent factor graphs. The transformation is illustrated in Fig. 2.17. In Bayesian networks, variable nodes with no parents have prior probabilities, and directed edges represent conditional probabilities. On the other hand, in factor graphs, factor nodes represent prior and conditional probabilities, and undirected edges represent only the connectivity.

Loopy max-sum is applied to the transformed factor graph. Let x_i be ith symbol in prototype trees, and θ and G be a set of parameters and Bayesian networks, respectively. Then individuals and MPS are represented by the following equations:

$$\mathbf{x} = (x_1, x_2, \ldots) \tag{2.94}$$

$$\mathbf{x}_{MPS} = \text{argmax}_{\mathbf{x}} P(\mathbf{x}; \theta, G). \tag{2.95}$$

Let $\mu_{f \to x}$ and $\mu_{x \to f}$ be messages from factor nodes to variable nodes and those from variable nodes to factor nodes, respectively, and $ne(X)$ and $ne(X) \backslash Y$ be the

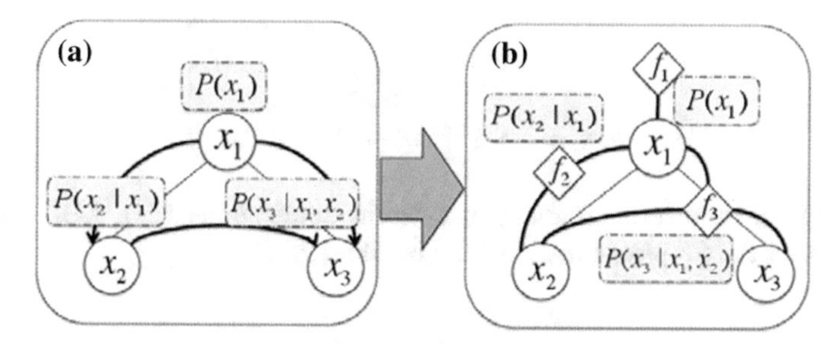

Fig. 2.17 **a** Bayesian network and **b** Factor graph. Dotted lines represent tree structures. Circles are variable nodes and lozenges are factor nodes. **b** is equivalent to **a**

set of adjacent nodes to X and the set of adjacent nodes to X except Y, respectively. $f(x_0, \ldots, x_n)$ represents the value of factor f when its adjacent variable nodes are $x_0, \ldots, x_n$. The details of loopy max-sum are described next.

Step 1 Initialization
All messages are initialized to 0.

Step 2 Message passing
Message passing is repeatedly executed, using Eqs. (2.96)–(2.99). Equations (2.96) and (2.97) represent messages from leaf nodes, and Eqs. (2.98) and (2.99) represent messages from nodes except leafs.

$$\mu_{f \to x}(x) = \ln f(x), \tag{2.96}$$

$$\mu_{x \to f}(x) = 0, \tag{2.97}$$

$$\mu_{f \to x}(x) = \max_{x_1, \ldots, x_N \equiv ne(f) \backslash x} \left[\ln f(x, x_1, \ldots, x_N) + \sum_{m \in ne(f_s) \backslash x} \mu_{x_m \to f}(x_m) \right], \tag{2.98}$$

$$\mu_{x \to f}(x) = \alpha_{xf} + \sum_{l \in ne(x) \backslash f} \mu_{f_l \to x}(x), \tag{2.99}$$

where α_{xf} is a normalization constant chosen such that

$$\sum_{x_n} \mu_{x \to f}(x) = 0.$$

Step 3 Check termination criteria
If termination criterion is satisfied, message passing terminates. Otherwise, message passing continues.

Step 4 Get the most probable solution (MPS)
Get the MPS, using Eqs. (2.100) and (2.101). x_i^{max} is the most probable instance of an ith and calculated from messages around an ith variable node.

$$x_i^{max} = \underset{x_i}{\operatorname{argmax}} \left[\sum_{s \in ne(x_i)} \mu_{f_s \to x_i}(x_i) \right] \tag{2.100}$$

$$\mathbf{x}_{MPS} \approx \mathbf{x}^{max} = (x_1^{max}, x_2^{max}, \ldots) \tag{2.101}$$

LBP works successfully in many real-world applications [21]. LBP has been employed for the for probabilistic model building in GP [62].

References

1. Aha, D.W., Kibler, D., Albert, M.: Instance-based learning algorithms. Mach. Learn. **6**, 37–66 (1991)
2. Angeline, P.J.: Evolutionary optimization versus particle swarm optimization: philosophy and performance differences. In: Porto, V.W., Saravanan, N., Waagen, D., Eiben, A.E. (eds.) Evolutionary Programming VII, pp. 601–610. Springer, Berlin (1998)
3. Bäck, T., Fogel, D.B., Michalewicz, Z. (eds.): Evolutionary Computation 1: Basic Algorithms and Operators. Institute of Physics Publishing, Bristol (2000)
4. Bishop, C.M.: Pattern Recognition and Machine Learning. Springer, Berlin (2006)
5. Buchberger, B.: A criterion for detecting unnecessary reductions in the construction of Gröbner-bases. In: Proceedings of the International Symposium on Symbolic and Algebraic Computation (EUROSAM'79), pp. 3–21 (1979)
6. Buchberger, B.: A note on the complexity of constructing Gröbner-bases. In: van Hulzen, J.A. (ed.) Computer Algebra, EUROCAL 1983. Lecture Notes in Computer Science, vol. 162. Springer, Berlin (1983)
7. Buchberger, B.: A critical pair completion algorithm for finitely generated ideals in rings. In: Borger, E., Hasenjaeger, G., Rodding, D. (eds.) Logic and Machines: Decision Problems and Complexity. Lecture Notes in Computer Science, vol. 171. Springer, Berlin (1984)
8. Buchberger, B.: Gröbner bases: an algorithmic method in polynomial ideal theory. In: Bose, N.K. (ed.) Multidimensional Systems Theory. D. Reidel Publishing Company, Dordrecht (1985)
9. Buchberger, B.: Applications of Gröbner bases in non-linear computational geometry. In: Jansen, R. (ed.) Trends in Computer Algebra. Lecture Notes in Computer Science, vol. 296. Springer, Berlin (1987)
10. Buchberger, B., Winkler, F.: Gröbner Bases and Applications, vol. 251. Cambridge University Press, Cambridge (1998)
11. Caviness, B.F., Johnson, J.R. (eds.): Quantifier Elimination and Cylindrical Algebraic Decomposition. Springer, Berlin (2013)
12. Chakraborti, N., Misra, K., Bhatt, P., Barman, N., Prasad, R.: Tight-binding calculations of Si-H clusters using genetic algorithms and related techniques: studies using differential evolution. J. Phase Equilibria **22**(5), 525–530 (2001)
13. Chang, C.-C., Lin, C.-J.: LIBSVM: a library for support vector machines. ACM Trans. Intell. Syst. Technol. **2**(3), 1–27 (2011)
14. Chou, S., Schelter, W.F.: Proving geometry theorems with rewrite rules. J. Autom. Reason. **2**, 253–273 (1986)
15. Cortes, C., Vapnik, V.: Support-vector networks. Mach. Learn. **20**(3), 273–297 (1995)
16. Dasarathy, B.: Nearest Neighbor (NN) Norms: NN Pattern Classification Techniques. IEEE Computer Society Press, Los Alamitos (1991)
17. Drucker, H.: Improving Regression Using Boosting Techniques. In: Proceedings of International Conference on Machine Learning (ICML97) (1997)
18. Dueck, D., Frey, B.J.: Non-metric affinity propagation for unsupervised image categorization. In: 2007 IEEE 11th International Conference on Computer Vision, pp. 1–8. IEEE (2007)

19. Eberhart, R.C., Shi, Y.: Comparison between genetic algorithms and particle swarm optimization. In: Proceedings of the Seventh Annual Conference on Evolutionary Programming, pp. 611–619 (1998)
20. Fan, R.-E., Chang, L.-W., Hsieh, C.-J., Wang, X.-R., Lin, C.-J.: LIBLINEAR: a library for large linear classification. J. Mach. Learn. Res. **9**, 1871–1874 (2008)
21. Felzenszwalb, P.F., Huttenlocher, D.P.: Efficient belief propagation for early vision. Int. J. Comput. Vis. **70**, 41–54 (2006)
22. Freund, Y., Schapire, R.E.: Experiments with a new boosting algorithm. In: Proceedings of International Conference on Machine Learning (ICML96) (1996)
23. Frey, B.J., Dueck, D.: Clustering by passing messages between data points. Science **315**, 972–976 (2007)
24. Fukushima, K.: Neocognitron: a self organizing neural network model for a mechanism of pattern recognition unaffected by shift in position. Biol. Cybern. **36**(4), 193–202 (1980)
25. Gämperle, R., Müller, S.D., Koumoutsakos, P.: A parameter study for differential evolution. In: Proceedings of International Conference on Advances in Intelligent Systems, Fuzzy Systems, Evolutionary Computation, pp. 293–298 (2002)
26. Goodfellow, I., Pouget-Abadie, J., Mirza, M., Xu, B., Warde-Farley, D., Ozair, S., Bengio, Y.: Generative adversarial nets. In: Advances in Neural Information Processing Systems, pp. 2672–2680 (2014)
27. Higashi, N., Iba, H.: Particle swarm optimization with Gaussian mutation. In: Proceedings of IEEE Swarm Intelligence Symposium (SIS03), pp. 72–79 (2003)
28. Hinton, G.E., Salakhutdinov, R.R.: Reducing the dimensionality of data with neural networks. Science **313**, 504–507 (2006)
29. Holland, J.H.: Adaptation in Natural and Artificial Systems. University of Michigan Press (1975)
30. Hu, R., Namee, B.M., Delany, S.J.: Off to a good start: using clustering to select the initial training set in active learning. In: Proceedings of the Florida Artificial Intelligence Research Society Conference (FLAIRS) (2010)
31. Huang, C.-H., Wang, C.-J.: A GA-based feature selection and parameters optimization for support vector machines. Expert Syst. Appl. **31**, 231–240 (2006)
32. Iba, H., Hirochika Inoue, H.: Reasoning of geometric concepts based on algebraic constraint-directed method. In: Proceedings of the IJCAI, pp. 143–151 (1991)
33. Iba, H., Noman, N.: New Frontiers in Evolutionary Algorithms: Theory and Applications. World Scientific Publishing Company, Singapore (2011)
34. Jia, Y., Wang, J., Zhang, C., Hua, X.-S.: Finding image exemplars using fast sparse affinity propagation. In: Proceedings of the 16th ACM International Conference on Multimedia, pp. 639–642. ACM (2008)
35. Kapur, D.: Geometric reasoning and artificial intelligence: introduction to the special volume. Artif. Intell. **37** (1988)
36. Kennedy, J., Eberhart, R.C.: Particle swarm optimization. In: Proceedings of IEEE the International Conference on Neural Networks, pp. 1942–1948 (1995)
37. Kennedy, J., Eberhart, R.C.: Swarm Intelligence. Morgan Kaufmann Publishers, San Francisco (2001)
38. Kennedy, J., Spears, W.M.: Matching algorithms to problems: an experimental test of the particle swarm and some genetic algorithms on the multimodal problem generator. In: Proceedings of the IEEE Congress on Evolutionary Computation (CEC), pp. 78–83 (1998)
39. Krink, T., Filipič, B., Fogel, G., Thomsen, R.: Noisy optimization problems – a particular challenge for differential evolution? In: Proceedings of Congress on Evolutionary Computation, pp. 332–339 (2004)
40. Krizhevsky, A., Sutskerver, I., Hinton, G.E.: ImageNet classification with deep convolutional neural networks. In: Advances in Neural Information Processing Systems 25 (NIPS), pp. 1097–1105 (2012)
41. Larsen, A.B.L., Sϕnderby, S.K., Winther, O.: Autoencoding beyond pixels using a learned similarity metric (2015). arXiv:1512.09300

42. Laubenbacher, R., Stigler, B.: A computational algebra approach to the reverse engineering of gene regulatory networks. J. Theor. Biol. **229**, 523–537 (2004)
43. Le, Q., Ranzato, M., Monga, R., Devin, M., Chen, K., Corrado, G., Dean, J., Ng, A.: Building high-level features using large scale unsupervised learning. In: Proceedings of the 29th International Conference on Machine Learning (2012)
44. LeCun, Y., Bottou, L., Bengio, Y., Haffner, P.: Gradient-based learning applied to document recognition. Proc. IEEE **86**(11), 2278–2324 (1998)
45. Limbeck, J.: Computation of approximate border bases and applications, Ph.D. thesis, Passau, Universität Passau, Dissertation (2014)
46. Liu, Z., Li, P., Zheng, Y., Sun, M.: Clustering to find exemplar terms for key phrase extraction. In: Proceedings of the 2009 Conference on Empirical Methods in Natural Language Processing, vol. 1, pp. 257–266. Association for Computational Linguistics (2009)
47. Loos, R.: Introduction. In: Buchberger, B., et al. (eds.) Computer Algebra Symbolic and Algebraic Computation. Springer, Berlin (1982)
48. Maclin, R., Opitz, D.: An empirical evaluation of bagging and boosting. In: Proceedings of National Conference on Artificial Intelligence (AAAI97) (1997)
49. Möller, H.M., Buchberger, B.: The construction of multivariate polynomials with preassigned zeros. In: Proceedings of the European Computer Algebra Conference on Computer Algebra, pp. 24–31 (1982)
50. Mourrain, B.: A new criterion for normal form algorithms. Applied Algebra, Algebraic Algorithms and Error-correcting Codes, pp. 430–442. Springer, Berlin (1999)
51. Nguyen, A., Yosinski, J., Clune, J.: Deep neural networks are easily fooled: high confidence predictions for unrecognizable images. In: Proceedings of 2015 IEEE Conference on Computer Vision and Pattern Recognition (CVPR), pp. 427–436 (2015)
52. Noman, N., Iba, H.: Enhancing differential evolution performance with local search for high dimensional function optimization. In: Proceedings of Genetic and Evolutionary Computation Conference (GECCO2005), pp. 967–974 (2005)
53. Noman, N., Iba, H.: A new generation alternation model for differential evolution. In: Proceedings of Genetic and Evolutionary Computation Conference (GECCO 2006), pp. 1265–1272 (2006)
54. Noman, N., Iba, H.: Differential evolution for economic load dispatch problems. Elsevier Electric Power Syst. Res. **78**(8), 1322–1331 (2008)
55. Noman, N., Iba, H.: Accelerating differential evolution using an adaptive local search. IEEE Trans. Evol. Comput. **12**(1), 107–125 (2008)
56. Pan, S.J., Yang, Q.: A survey on transfer learning. IEEE Trans. Knowl. Data Eng. **22**(10), 1345–1359 (2010)
57. Paul, T.K., Ueno, K., Iwata, K., Hayashi, T., Honda, N.: Genetic algorithm based methods for identification of health risk factors aimed at preventing metabolic syndrome. In: Proceedings of the 7th International Conference on Simulated Evolution And Learning (SEAL'08). LNCS, vol. 5361, pp. 210–219. Springer, Berlin (2008)
58. Pearl, J.: Probabilistic Reasoning in Intelligent Systems: Networks of Plausible Inference. Morgan Kaufmann, San Mateo (1988)
59. Price, K.V., Storn, R.M., Lampinen, J.A.: Differential Evolution: A Practical Approach to Global Optimization. Springer, Berlin (2005)
60. Quinlan, J.R.: Bagging, Boosting, and C4.5. In: Proceedings of National Conference on Artificial Intelligence (AAAI96) (1996)
61. Radford, A., Metz, L., Chintala, S.: Unsupervised representation learning with deep convolutional generative adversarial networks (2015). arXiv:1511.06434
62. Sato, H., Hasegawa, Y., Bollegala, D., Iba, H.: Improved sampling using loopy belief propagation for probabilistic model building genetic programming. Swarm Evol. Comput. **23**, 1–10 (2015)
63. Stifter, S.: Algebraic methods for computing inverse kinematics. J. Intell. Robot. Syst. **11**(1–2), 79–89 (1994)

64. Storn, R.: System design by constraint adaptation and differential evolution. IEEE Trans. Evol. Comput. **3**(1), 22–34 (1999)
65. Storn, R., Price, K.V.: Differential evolution – a simple and efficient adaptive scheme for global optimization over continuous spaces. Technical report TR-95-012, ICSI (1995)
66. Storn, R., Price, K.V.: Differential evolution -a simple and efficient heuristic for global optimization over continuous spaces. J. Global Optim. **11**(4), 341–359 (1997)
67. Swan, J., Neumann, G.K., Krawiec, K.: Analysis of semantic building blocks via Grobner bases. In: Johnson, C., Krawiec, K., O'Neill, M., Moraglio, A. (eds.) Semantic Methods in Genetic Programming (SMGP) at Parallel Problem Solving from Nature (PPSN XIV), Ljubljana, Slovenia (2014)
68. Tipping, M.E.: The relevance vector machine. Advances in Neural Information Processing Systems, pp. 652–658. MIT Press, Cambridge (2000)
69. Vapnik, V.: Statistical Learning Theory. Wiley, New York (1998)
70. Yang, Y., Iba, H.: Fooling voice based on evolutionary computation. In: Proceedings of Evolutionary Computation Symposium, Dec. 9–10, Hokkaido, Japan (2017)
71. Zaharie, D.: Critical values for the control parameters of differential evolution algorithms. In: Proceedings of MENDEL 2002, 8th International Conference on Soft Computing, pp. 62–67 (2002)
72. Zha, Z.-J., Yang, L., Mei, T., Wang, M., Wang, Z.: Visual query suggestion. In: Proceedings of the 17th ACM International Conference on Multimedia, pp. 15–24. ACM (2009)

Chapter 3
Evolutionary Approach to Deep Learning

A sonnet written by a machine would be better appreciated by another machine.

(Alan Turing)

Abstract This chapter describes an evolutionary approach to deep learning networks. We first explain neuroevolution approach, which can adaptively learn a network structure and size appropriate to the task. A typical example of neuroevolution is NEAT. NEAT has demonstrated performance superior to that of conventional methods in a large number of problems. Then, several studies on deep neural networks with evolutionary optimization are explained, such as Genetic CNNs, hierarchical feature construction using GP, and Differentiable pattern-producing network (DPPSN).

Keywords Neuroevolution · Neuroevolution of augmenting topologies (NEAT) HyperNEAT · L–system · Composition pattern-producing network (CPPN)

3.1 Neuroevolution

"Neuroevolution" is an active area of research in machine learning. This approach, also known as evolutionary artificial neural networks (EANNs) [22], integrates evolutionary methods with neural networks. The key feature of evolutionary neural networks is a genetic search for an optimal network. This saves the effort normally required to search for a neural network (such as constructing networks through trial and error).

Training via backpropagation, the method normally used to train neural networks, has been shown to often become stuck in only locally optimal solutions because it is based on the gradient descent method. To address this flaw, methods have been proposed that use evolutionary methods to learn which link weights to use. In other words, these methods express the link weights in a network as genotypes and then

H. Iba, *Evolutionary Approach to Machine Learning and Deep Neural Networks*, https://doi.org/10.1007/978-981-13-0200-8_3

use evolutionary computation to search for the best network. In this case, we derive the fitness value by the error in the output of the neural network (represented as a genotype, with lower errors representing better fitness). Methods for expressing link weights (genotypes) include binary strings and numeric vectors.

Note also that, when training a neural network, the network structure needs to be provided in advance. Neuroevolution, in contrast, can adaptively learn a network structure and size (number of nodes) appropriate to the task. Two different genotype representations have been proposed for driving the evolution of network structures.

Direct coding Directly represents the link state of the network structure.
Indirect coding Encodes production rules for generating networks as genotypes.

Indirect coding is closer to biological models than direct coding is. For example, direct coding might represent a network consisting of N nodes $(n_1, n_2, \ldots n_N)$ as an $N \times N$ adjacency matrix, with each element taking a value of 0 or 1 such that a value of 1 for row i and column j indicates that there is a link from n_i to n_j and a value of 0 indicates that there is no such link.

For indirect coding methods, the focus of attention has been encoding for developmental systems. This approach uses observations from developmental biology (see Sect. 5.1) as a basis to create abstract models of natural development in order to drive evolutionary computation algorithms. These models range from low-level cytochemistry through to high-level grammar-rewriting systems.

One example of a high-level method is the use of L-systems, mathematical models introduced by Lindenmayer in the 1960s to describe interactions between cells during ontogenesis [12]. These models describe morphological mutations through the application of rewrite rules. Consider the following rewrite rule, for example.

```
F -> F[+F][-F]
```

If we assume that F is the initial symbol then the following strings will be produced in the first three iterations.

1. `F[+F][-F]`
2. `F[+F][-F][+F[+F][-F]][-F[+F][-F]]`
3. `F[+F][-F][+F[+F][-F]][-F[+F][-F]][+F[+F][-F][+F[+F][-F]] [-F[+F][-F]]][-F[+F][-F][+F[+F][-F]][-F[+F][-F]]]`

Here we take F to be a branch, with square brackets [and] indicating branch bifurcations and + and − indicating growth of the left (+36°) or right (−36°), respectively. These strings then represent the growth process of trees, as shown in Fig. 3.1. Adding even more complex rules allows us to render even more realistic trees (Fig. 3.2). For example, Fig. 3.2b uses the rewrite rule below, except that for this rule the + and − rotations are now 25°.

```
F -> FF+[+F-F+F]-[-F+F-F]
```

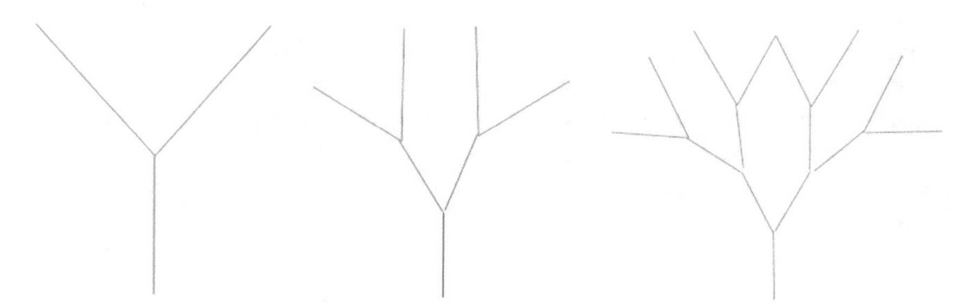

Fig. 3.1 Growth process of trees

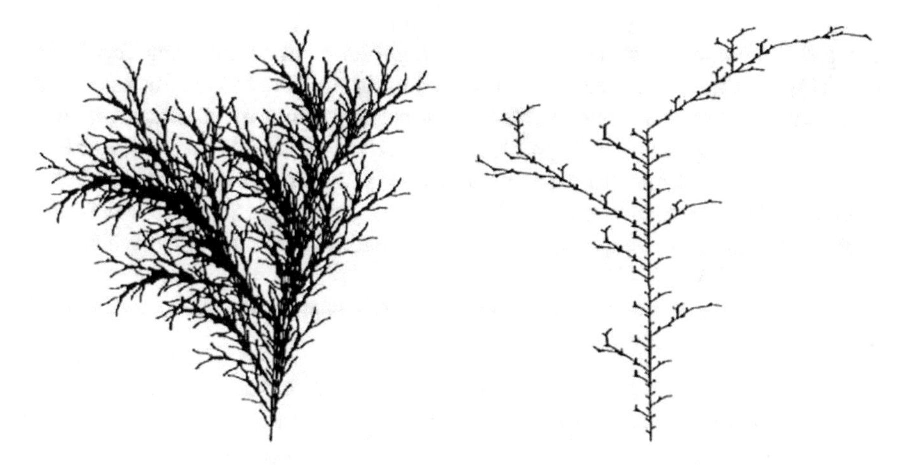

Fig. 3.2 Generated trees by L–systems

L–systems have been actively adopted for applications such as computer animation due to the ease with which the growth processes of living organisms can be described. This makes it possible to obtain extremely realistic images that one would never imagine had been drawn by a computer.

Once the genotype is determined, the following procedure is used to conduct training through evolutionary computations on network structures.

Step1 Find the phenotypes (the network structure of the EANNs) by decoding the genotypes for each individual.
Step2 Train each EANN. In this case, change the weights according to a predetermined training rule (such as backpropagation).
Step3 Derive the fitness value for each individual based on the training results. Normally, the squared error relative to the desired output is used for this.
Step4 Select candidate parents according to the fitness values, and then create the child generation by applying genetic operators to the selected parents.
Step5 Return to **Step1** if the termination condition has not been met.

Furthermore, although Hebb's rule is the training rule typically used for neural networks, parameters such as the learning rate and momentum during backpropagation are not easy to tune because they are task dependent. Accordingly, there is active research into evolutionary methods for learning appropriate parameters as well as into using evolutionary computation to search for more general training rules than Hebb's rule.

3.1.1 NEAT and HyperNEAT

A typical example of neuroevolution is NeuroEvolution of Augmenting Topologies (NEAT) [16], a method for efficiently optimizing the structure and parameters of neural networks. NEAT has demonstrated performance superior to that of conventional methods in a large number of problems. The NEAT method makes it possible to evolve variable network structures by gradually increasing the complexity of a low number of small networks with each successive generation. Section 5.4 describes neuroevolution methodologies and specific applications based on ERNe, which is an extension of NEAT.

Figure 3.3 shows how NEAT produces network phenotypes from genotypes. A genotype consists of two parts.

- Node information: lists of input nodes, hidden nodes, and output nodes.
- Link information: a list of the arcs linking pairs of nodes.

Link information contains descriptions of the nodes forming the beginning and end points of an arc, link weightings, flags indicating whether an arc is enabled or disabled, and ID numbers. For example, in Fig. 3.3 there are three input nodes, one hidden node and one output node. Note also that although seven arcs are described in the link information, the second arc is disabled and therefore not expressed in the phenotype (the corresponding arc does not exist in the network).

NEAT includes two types of spontaneous mutations (see Fig. 3.3).

- Node additions.
- Arc additions.

In Fig. 3.3, the arrays describe the link information within genes. The numbers at the top of each gene are ID numbers. New genes are allocated ID numbers in such a way that the sequence of ID numbers is strictly increasing. In the top example in Fig. 3.4, an arc has been added. Here the new gene (arc) has been assigned the next available ID number (in this case, 7). In the bottom example, the arc to be removed is first disabled, and then two new arc genes (with IDs 8 and 9) are added. A new node is added between these two arcs, and the node information is recorded as a new gene (the node information descriptions have been omitted from the figure).

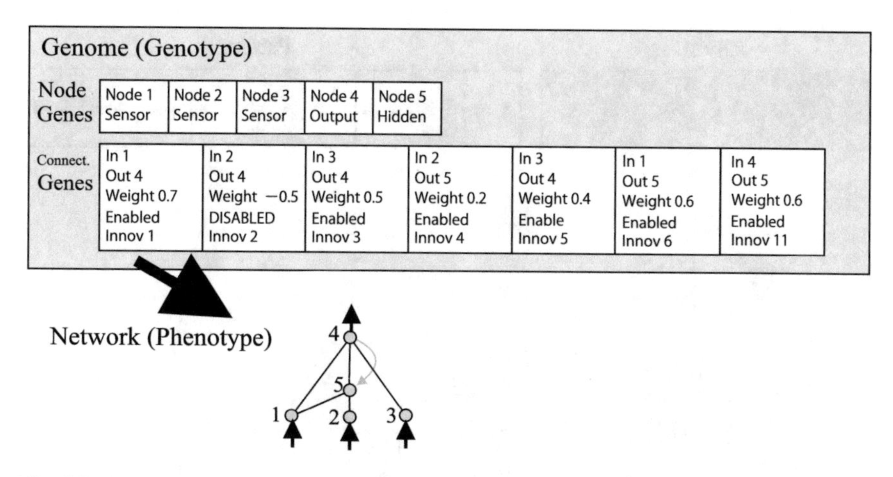

Fig. 3.3 A genotype to phenotype mapping example (Reproduced from [16])

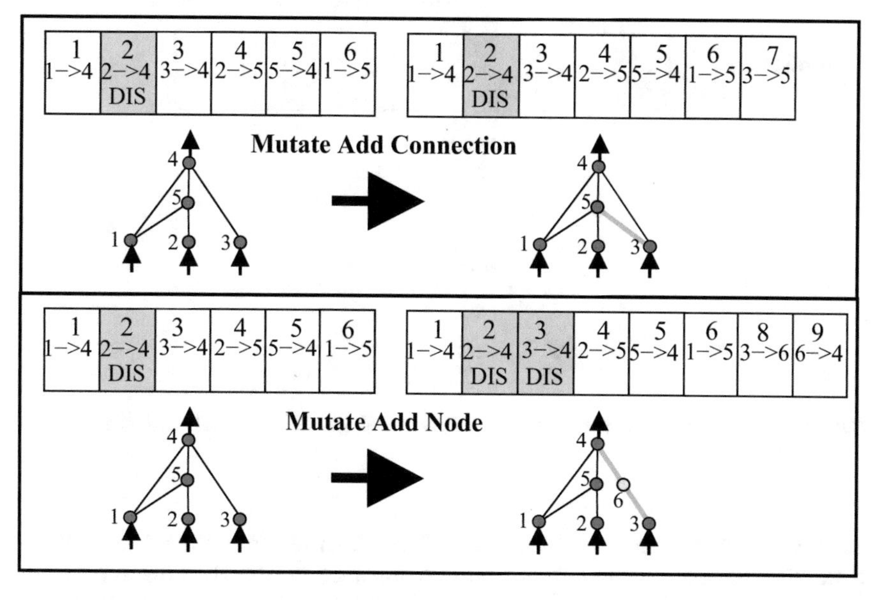

Fig. 3.4 The two types of structural mutation in NEAT (Reproduced from [16])

When genes are crossed, children inherit their parents' genes and retain the same ID numbers. In other words, the ID numbers do not change. In this way, the representation makes it possible to track the history (ancestry) of each gene.

Figure 3.5 shows how genetic crossover works with NEAT. Parent 1 and Parent 2 have different shapes, but by looking at the commonality between the gene ID numbers we can tell which genes match. Matching genes are inherited at random. In contrast, disjoint genes (the non-matching genes in the middle section) and excess

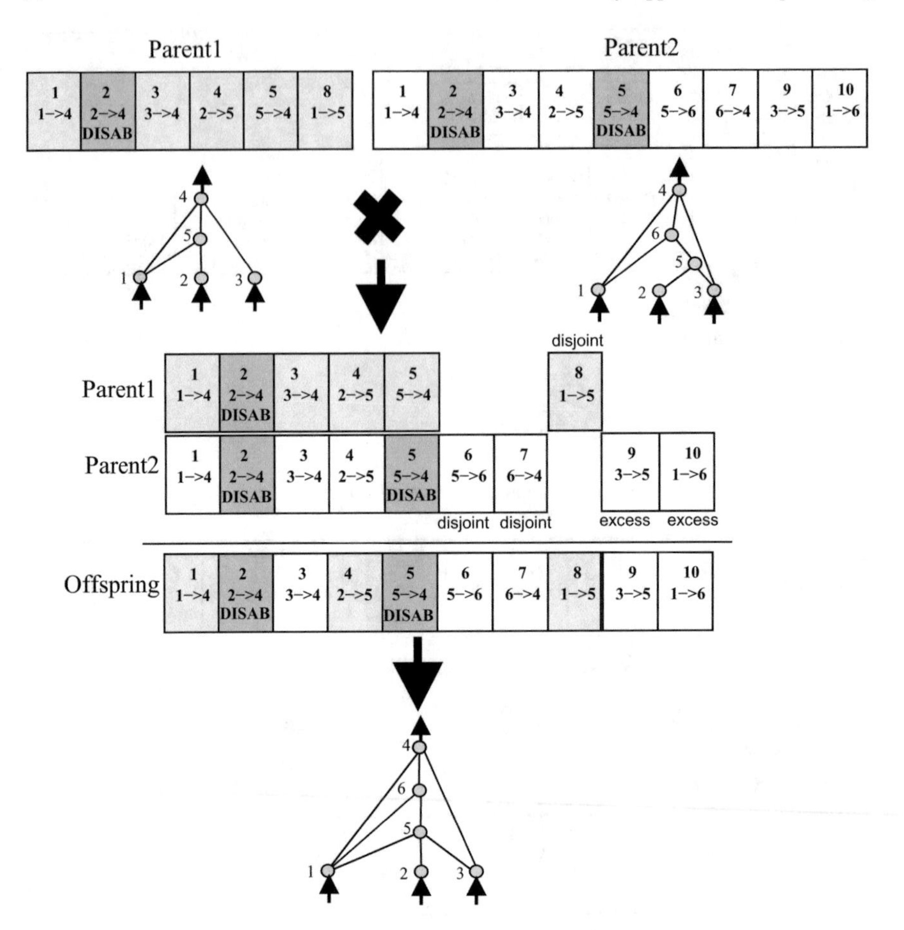

Fig. 3.5 Matching up genomes for different network topologies using innovation numbers (Reproduced from [16])

genes (the non-matching genes at the end) are inherited from the parent with the better fitness. In the figure, both parents are assumed to have the same fitness, and so genes are inherited randomly in both cases. Disabled (resp., enabled) genes may become enabled (disabled) in the future.

The efficacy of NEAT has been verified through application to a wide variety of fields. For example, modular NEAT has demonstrated better results than conventional methods in learning to play the Ms. Pac-Man video game (Fig. 3.6) [13]. This NEAT expresses multiple modules simultaneously. Each module learns different policies, and decisions as to which policy to use at any given time are also acquired through evolutionary processes. Inputs take the form of feature values, such as the positions of the player and the ghosts. In games such as Pac-Man, where rewards and penalties are delayed and so conditions need to be judged from a long-term perspective, it has also been pointed out that traditional reinforcement learning does not perform

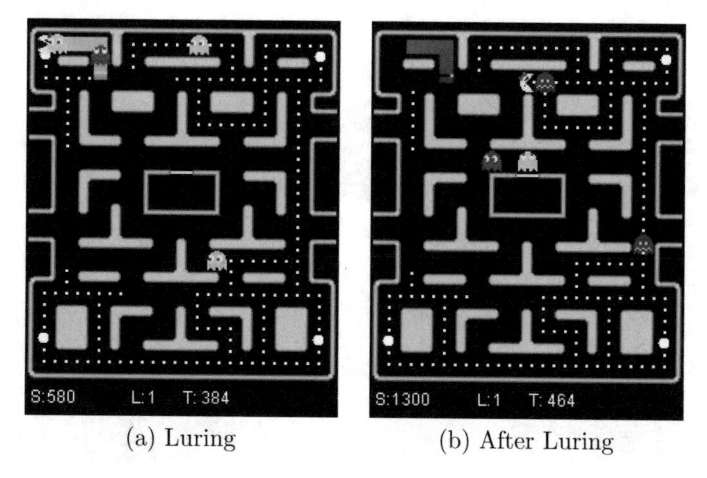

(a) Luring (b) After Luring

Fig. 3.6 Behavior with a luring module (Reproduced from [13])

adequately. Modular NEAT succeeds in scoring points by training "luring modules" and appropriately selecting which one to use according to the situation (Fig. 3.6).

3.1.2 CPPN and Pattern Generation

NEAT has been extended via a coding method known as composition pattern-producing network (CPPN) to create hyperNEAT, which has been the focus of much attention.

CPPN is a model for developmental systems that can express complex recursive patterns by using Cartesian coordinates. Fundamentally, this is a network in which coordinate values are received as inputs and the phenotypes at the named location are acquired as outputs.

Figure 3.7 shows a schematic of the behavior in the two-dimensional case. In Fig. 3.7a, the genotype is a function f and the corresponding phenotype is a spatial pattern generated by passing two-dimensional coordinates (x, y) as inputs to f and then using the resulting value at the (x, y) coordinate position as the output. To express this function f, CPPN uses encoding via a network formed by composing various functions. Figure 3.7b shows a two-dimensional CPPN in which the input is the coordinate (x, y) and the output is the value of the phenotype obtained at that coordinate. Just as for neural networks, weights are defined for each link, and the output from a given node is multiplied by the link weight before becoming the input value for the next node.

CPPN can use primitive functions (such as Gaussian and sine functions) to express complex regularities and symmetries, as shown in Fig 3.8. Note that this figure was arrived at via interactive evolutionary computations (IEC, see Sect. 1.6) [15]. The

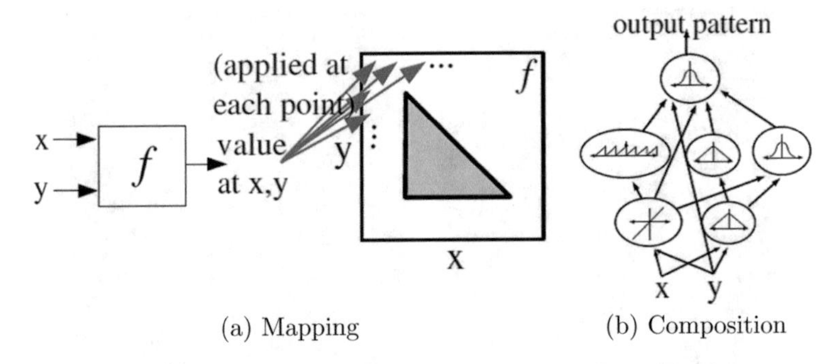

Fig. 3.7 CPPN Encoding (Reproduced from [15])

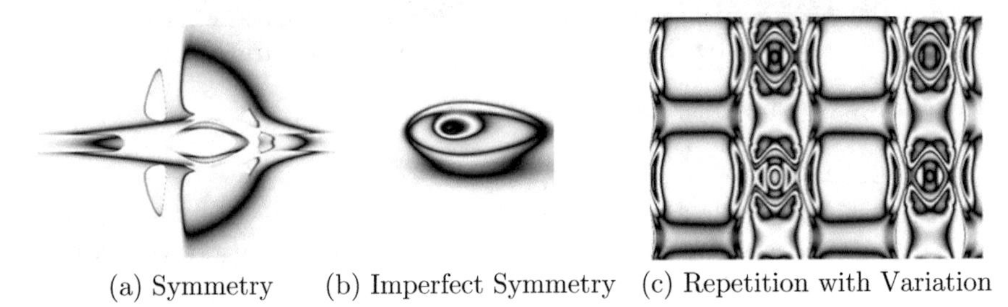

Fig. 3.8 CPPN-generated regularities (Reproduced from [15])

primitive functions used here model particular phenomena in developmental systems. For example, the Gaussian function models left–right symmetry, while periodic functions (such as the sine function) model differentiation into body segments.

The resulting phenotypes (Fig 3.8) are geometrically important features observed in the natural world. For example, the figure in (a) can be viewed as a model for bilateral symmetry in vertebrates, while the imperfect symmetry and recursion in (b) can be thought of as models for right-handedness and receptive fields in the cortex, respectively, and the recursion accompanied by transformation in (c) can be thought of as a model for the cortical column or columnar structures.[1]

There are other methods for generating spatial patterns similar to those produced by CPPN. One example of graphic art based on IEC is Sbart [20], a system for generating two-dimensional images using Genetic Programming. Sbart renders images in accordance with calculations on two-dimensional vectors including variables x and y. Substituting the x and y coordinates for each point (pixel) in the image into the formula and then appropriately converting the values obtained into rendering information results in strikingly beautiful two-dimensional images[2] (see Fig. 3.9).

[1] A columnar structure in the cerebrum where neurons with similar properties concentrate. In mouse perceptual fields, there are believed to be columns corresponding to each individual whisker.

[2] This image was created by LGPC for Art, a simulator created with reference to Sbart. This tool can be used to "nurture" to influence their creation. See our Web site for information on installing and using LGPC for Art.

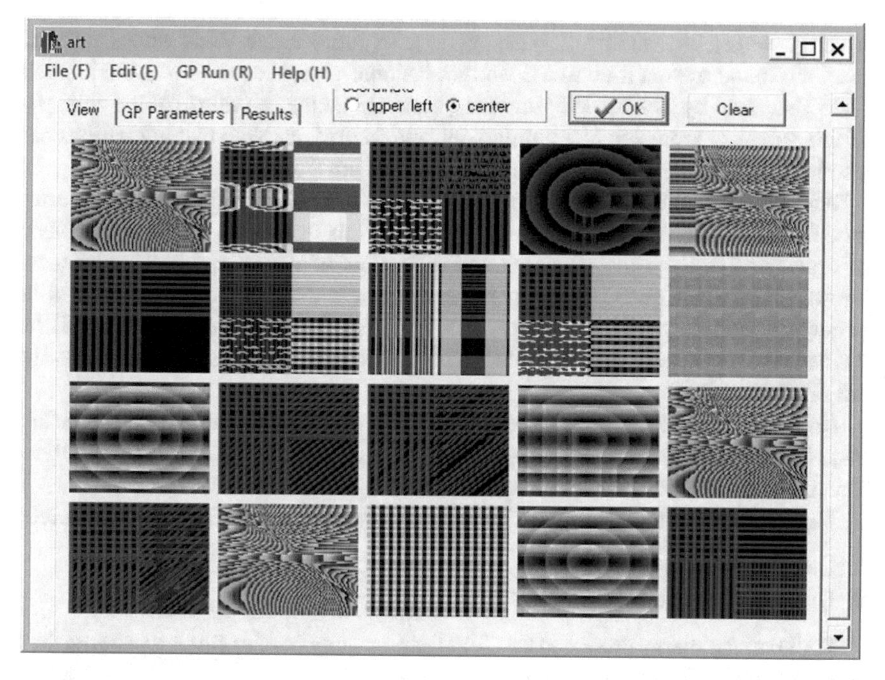

Fig. 3.9 LGPC for Art

For example, the genotype for the third diagram (phenotype) from the left in the top row of the figure is as follows[3]:

$$RGB = \min(\vec{x0y}, \min(\vec{yx0}, hypot(\vec{0xy}, \vec{0yx}))) - \max(5.400, \frac{\vec{xy0}}{\min(\vec{xy0}, \vec{yx0})}).$$

A three-dimensional vector consisting of three identical real values acts as a constant term, while a three-dimensional vector consisting of the image coordinates and the constant 0 acts as a variable term. In the formula above, $\vec{yx0} = (y, x, 0)$, $\vec{0xy} = (0, x, y)$ and so on. For the functions, values are calculated for each individual element.

Functions max and min perform processing by comparing the values of the first element in each vector. For example, if $x = (x_1, x_2, x_3)$ and $y = (y_1, y_2, y_3)$, then max is defined as follows:

$$\max(x, y) = \begin{cases} x & \text{When } x_1 > y_1 \\ y & \text{Otherwise,} \end{cases}$$

[3] Here $hypot$ gives the Euclidean distance on two dimensions, defined as $hypot(x, y) = \sqrt{x^2 + y^2}$.

where, $x = (x_1, x_2, x_3)$ and $y = (y_1, y_2, y_3)$. In Fig. 4.9, the RGB values for each pixel are found by first finding a three-dimensional vector based on a numerical formula generated by Genetic Programming and then taking the values of the elements of this vector to represent hue, luminosity, and saturation. Note that animations can also be generated by including a time variable where 0 appears.

Another method that is closer to biological pattern generation is based on Turing models (Fig. 3.10). This is a complex simulation that explains the morphogenesis of organisms through the reactions and diffusions of virtual chemical substances known as morphogens. By changing the generation and diffusion parameters, a wide variety of patterns can be generated, such as those shown in the figure (see [5] for the details). These patterns resemble the designs on a seashell or the patterns on the tail of a leopard.

Note the structural similarity between CPPN and neural networks. This means that NEAT can also be applied to optimizing the structure and parameters of CPPN. This method is known as hyperNEAT.

The basic idea of hyperNEAT [17] can be summarized as the following two points.

1. Use CPPN to express the links and weights of a neural network.
2. Use NEAT to optimize CPPN.

The spatial patterns produced by CPPN can be thought of as linking patterns for a neural network. For example, if a four-dimensional CPPN outputs w in response to the input (x_1, y_1, x_2, y_2), then this can be interpreted as meaning that the link weight between a node at coordinates (x_1, y_1) and another node at coordinates (x_2, y_2) is w. When the absolute value of w is below a certain value, this is taken to indicate that there is no link between the nodes (a weight indistinguishable from 0). Note that the output values for CPPN are scaled appropriately. Consider, for example, nodes placed on a 5×5 lattice, with a normal two-dimensional coordinate system with the origin located at the center of the lattice. Given these node positions, the link weights between pairs of nodes are found by passing their coordinates to the CPPN (Fig. 3.11). Recall the regular and symmetrical spatial patterns produced by CPPN. These kinds of patterns generate regular linkage patterns for the neural network. For example, CPPN can easily express the link relationships shown in Fig. 3.12, such as symmetry, incomplete symmetry, recursion, and recursion accompanied by transformation.

Moreover, the input and output layers of a neural network can be set to arbitrary locations. This enables more geometrical relationships to be used. For example, Fig. 3.13 shows input layers whereby a robot's input sensors (I) and output modules (O) are placed in either a circle or in parallel.

By precisely arranging neurons in this way, geometrical regularity can be achieved using CPPN encoding. Biological neural networks have many such functions. The neurons in the visual cortex are arranged in the same topological two-dimensional pattern as the light receptors in the retina, an arrangement that is believed to acquire locality through regular connections to adjacent neurons based on simple recursion. In a certain group of fish, various pyramidal cells are known to be arranged in regular arrays along the curved surface of the retina. For example, a pattern of four different

Fig. 3.10 Turing model simulation results

pyramidal cells can be observed in the retina of zebra fish, each with different peaks in light sensitivity, corresponding to the wavelengths for blue, red, green, and ultraviolet light.

CPPN also has the same kind of capacity for pattern composition. In fact, geometrically valid information provides a domain-specific bias in evolutionary processes that often surpasses simple optimization methods.

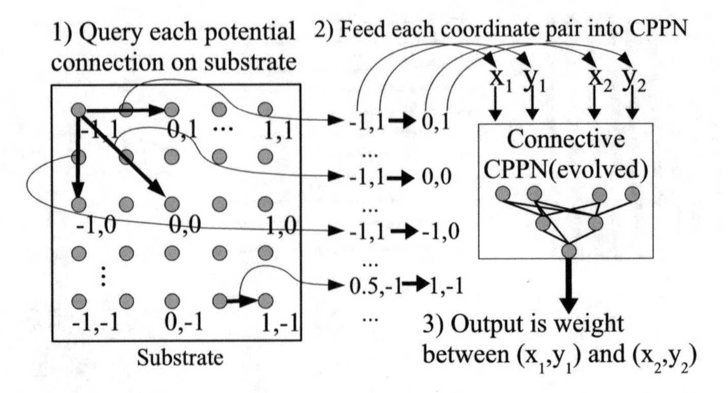

Fig. 3.11 Hypercube-based Geometric Connectivity Pattern Interpretation (Reproduced from [17])

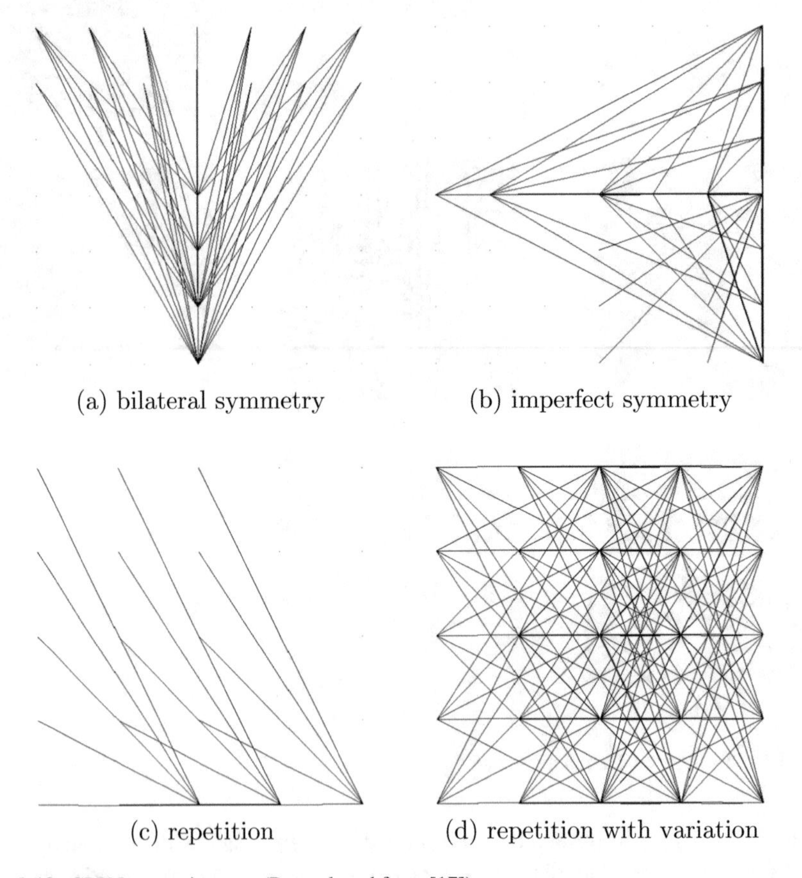

(a) bilateral symmetry (b) imperfect symmetry

(c) repetition (d) repetition with variation

Fig. 3.12 CPPN output images (Reproduced from [17])

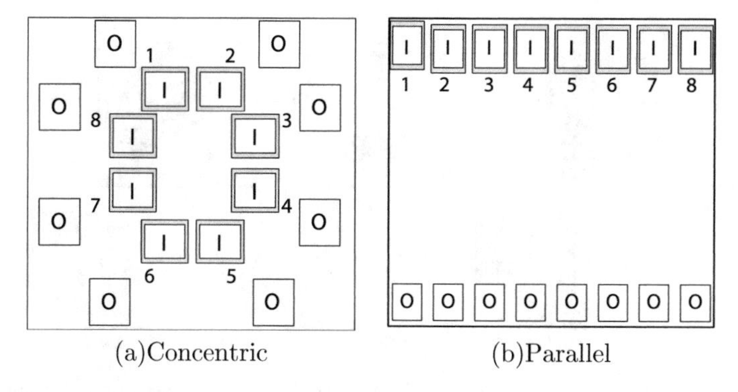

Fig. 3.13 Placing inputs and outputs (Reproduced from [17])

HyperNEAT has been applied to Atari games [2]. It has also been applied to video games such as Super Mario, and there is great interest in its efficacy.[4]

In recent years, DeepNEAT has been proposed, which is a most immediate extension of the standard neural network topology-evolution method NEAT to DNN [10].

3.2 Deep Neural Networks with Evolutionary Optimization

3.2.1 Genetic Convolutional Neural Networks (Genetic CNNs)

Xie et al. [21] performed a search of a deep neural network (DNN) using GAs. CNNs with a limited number of stages handle a network of predefined constituent elements, such as each stage being a convolution layer and pooling layer. Even with such limitations, the number of possible networks increases exponentially with the number of layers, complicating the search for the optimum network by simply counting them all.

The network handled here consists of S stages, and the sth stage consists of K_s nodes. The nodes within each stage are ordered, and it is assumed that there are only connections from a node with a small number to a node with a large number. Each node corresponds to a convolution operation, which is conducted after taking the sum of all the input nodes (i.e., nodes with smaller numbers) for each

[4]http://gigazine.net/news/20150616-mari-o/.

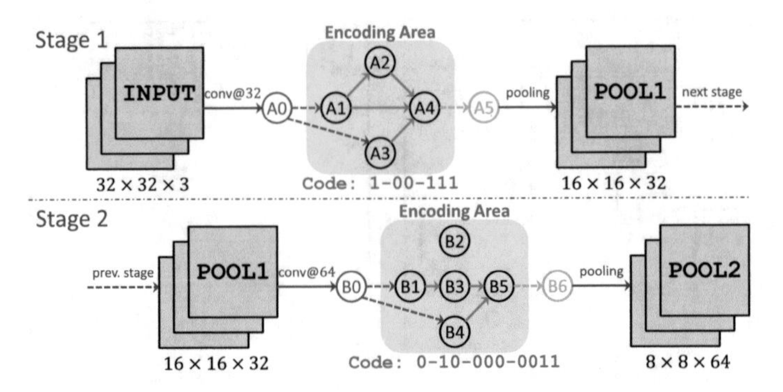

Fig. 3.14 Example of a genotype and a phenotype with a two-stage network (Reproduced from [21])

element. After convolution, batch normalization [6][5] and rectified linear unit (ReLU[6]) implementations are conducted. Note that the network handled here is not one comprised of all connections. Figure 3.14 shows an example of a two-stage network ($S = 2$, $(K1, K2) = (4, 5)$). The genotype is a binary string. For example, Stage 1 is the genotype

```
1-00-111
```

A0 and A5 are input and output nodes, respectively. The genotype encodes the connections between the intermediate nodes other than these input/output nodes. As the number of intermediate nodes at Stage 1 is 4, the genotype length is $3 \times 2 = 6$, which describes the adjacency relationships respectively as follows ($1 =$ has relationship, $0 =$ no relationship)

$$A1 \rightarrow A2 \mid A1 \rightarrow A3 \mid A2 \rightarrow A3 \mid A1 \rightarrow A4 \mid A2 \rightarrow A4 \mid A3 \rightarrow A4$$

In addition, an intermediate node with no input is connected such that there is an input from an input node, and an intermediate node with no output is connected such that there is output to an output node. The number of convolution filters at each stage is constant. In this example, there are 32 convolution filters at Stage 1 and 64 at Stage 2. Spatial resolution is unchanged (32×32 at stage 1, 16×16 at stage 2). Each pooling layer is down-sampled by a factor of 2.

In general, when the number of intermediate nodes of Stage s is K_s, the genotype length is $\frac{1}{2}K_s(K_s - 1)$, and each element indicates the existence or lack of an adjacency relationship as follows:

[5]During training, each layer's inputs are normalized across the current minibatch to the Gaussian distributions (usually zero mean and unit variance). It has been shown to have several benefits, e.g., faster convergence, easier to escape from local optima, more robust network.

[6]A function outputting 0 when the input is 0 or less and outputting the input as it is when the input is greater than 1.

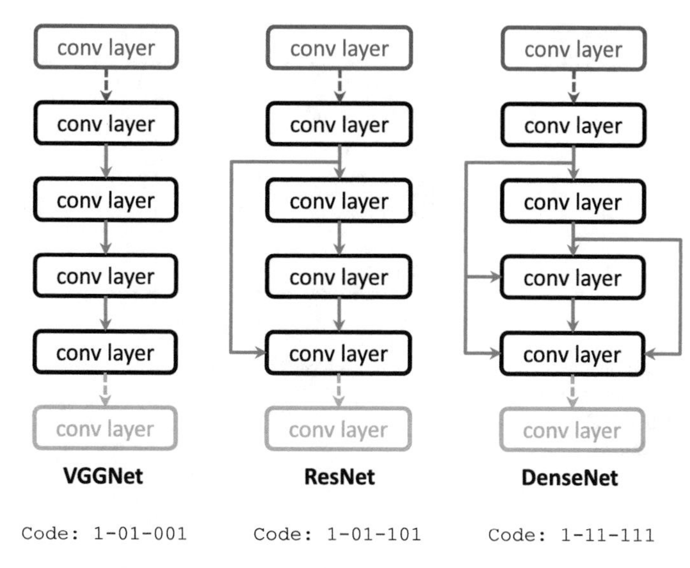

Fig. 3.15 Representative deep neural network genotypes (Reproduced from [21])

$$A_1 \rightarrow A_2 \mid A_1 \rightarrow A_3 \mid A_2 \rightarrow A_3 \mid \cdots \mid A_1 \rightarrow A_{K_s} \mid A_2 \rightarrow A_{K_s} \mid A_3 \rightarrow A_{K_s} \mid \cdots \; A_{K_s-1} \rightarrow A_{K_s}$$

This genotype can also represent VGGNet [14], ResNet [3], DenseNet [4], and so on, which are often used in DNN. Figure 3.15 shows these phenotypes and their genotypes. ResNet has a structure that adds residual functions between layers. Instead of learning the optimum output obtained in a certain layer, optimization is improved by learning the residual function referring to the input of the layer.

The CIFAR10 dataset [8] was used in learning. These data have 32 × 32 RGB images and consist of ten categories. A total of 50,000 images are used as training examples, with 10,000 images used as test examples. The population size for GA is 20, and the maximum number of generations is 50. The evaluation of each individual took 0.4 hours on average, and the total execution time was 17 GPU-days. Ten GPUs were used for the experiment.

Table 3.1 summarizes the results of the computational experiment. CIFAR10 is a classification problem including ten basic classes, while CIFAR100 is one that includes 100 classes. The SVHN data set [11] is larger in scale than that used for numerical recognition, with 73,257 training examples, 26,032 test examples, and 531,131 extra training examples. Each piece of data is a 32 × 32 RGB image. GeNet after G-N indicates the best individual in generation N. G-00 is the best individual from among randomly generated initial individuals. Additionally, GeNet#1 and GeNet#2 are the best individuals obtained as a result of evolution, and this structure is shown in Fig. 3.16. For comparison, GeNet from WRN is the result of a residual network [23].

In order to evaluate the obtained network, the ILSVRC 2012 classification task was executed (see p. 68). The network structure was fixed, and parameter adaptation

Table 3.1 Comparison of the recognition error rates (%)

	SVHN	CIFAR10	CIFAR100
GeNet after G-00	2.25	8.18	31.46
GeNet after G-05	2.15	7.67	30.17
GeNet after G-20	2.05	7.36	29.63
GeNet#1 (G-50)	1.99	7.19	29.03
GeNet#2 (G-50)	1.97	7.10	29.05
GeNet from WRN [23]	1.71	5.39	25.12

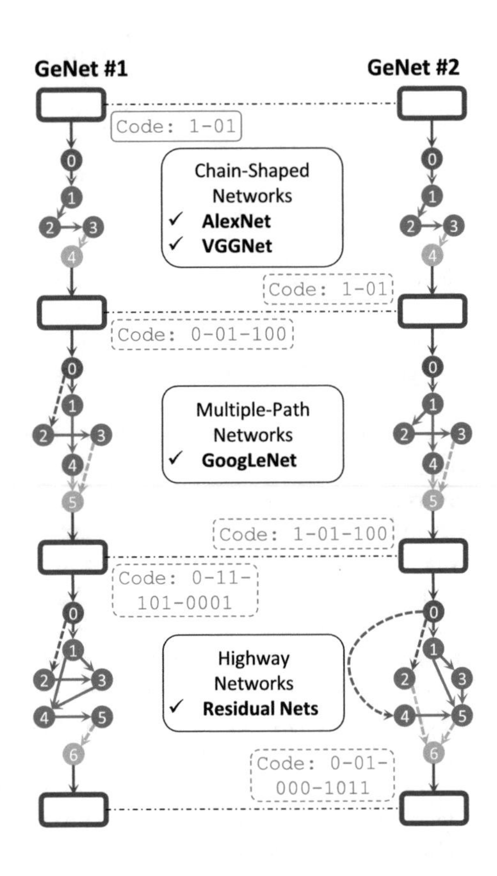

Fig. 3.16 Example of a genotype and a phenotype in a convolutional neural network (Reproduced from [21])

alone was conducted. Learning took approximately 20 GPU-days. In the first two stages, VGGNet (with four convolutional layers and two pooling layers) was applied and the dimension of data changed to $56 \times 56 \times 128$. Thereafter, classification was conducted by applying the network shown in Fig. 3.16.

Table 3.2 summarizes the results of the experiment, which shows that the structure obtained by learning from the small dataset (CIFAR10) is also good for the large-scale dataset (ILSVRC2012). The network obtained by evolution has yielded better results than VGGNet-16 and VGGNet-19. This is due to the fact that the original

Table 3.2 Top-1 and top-5 recognition error rates (%)

	Top-1	Top-5	# Paras.
AlexNet [9]	42.6	19.6	62M
GoogLeNet [19]	34.2	12.9	13M
VGGNet-16 [14]	28.5	9.9	138M
VGGNet-19 [14]	28.7	9.9	144M
GeNet#1	28.12	9.95	156M
GeNet#2	27.87	9.74	156M

chain structure (i.e., chain-styled stages) has been replaced by a structure of evolution that yields results proven to be more effective.

3.2.2 Hierarchical Feature Construction Using GP

Suganuma [18] et al. proposed a method to recognize feature construction by optimizing two types of processing by evolutionary computing:

- Image conversion by a combination of existing image processing filters.
- Image conversion by filter processing constructed by CGP.

Using this method, it is possible to reduce the solution space by dividing the optimization into two stages, rather than simultaneously optimizing the parameters of the entire multilayer structure as is done in CNN implementations.

The processing flow of the first stage of feature construction is as follows:

- Image conversion (filtering layer) using an existing image processing filter on the input image.
- Pooling processing (pooling layer).
- Classification (classification layer).

In the second stage of feature construction, image conversion is conducted via filter processing in which the converted image obtained in the first step is further constructed with CGP (see Sect. 1.5). As in the first step, a pooling process is performed on the image after conversion, and each pixel value after the pooling process is input to the classifier as a feature quantity. A filter is constructed by CGP such that the classification accuracy is high for the verification image set.

The CNN of the phenotype defined by the CGP genotype is trained with the training dataset, and the results of the validation set are derived as the fitness value of the genotype. Next, the more suitable CNN structure is searched through the evolution.

Figure 3.17 shows an example of a genotype and a phenotype. The left genotype defines the architecture of the CNN on the right. In this case, node No.5 in the diagram on the left is inactive. The function node for CGP used here is shown in Table 3.3.

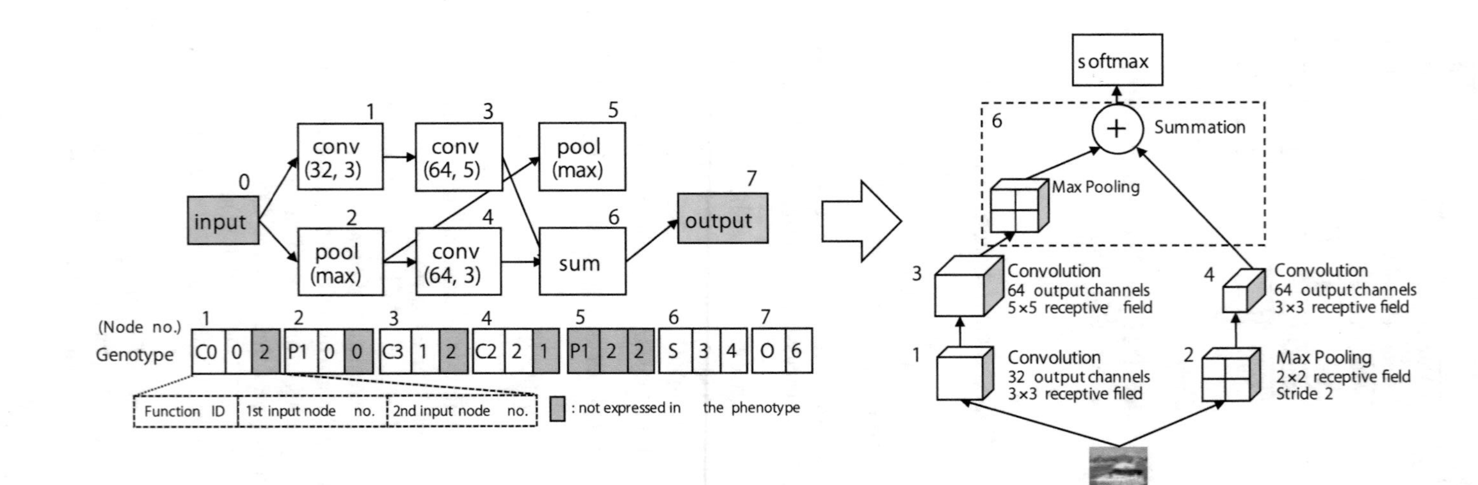

Fig. 3.17 Example of a genotype and a phenotype (Reproduced from [18])

Table 3.3 Node functions in CGP(C': number of output channels, k: receptive field size or kernel size)

Node type	Symbol	Variation
Convolution block	CB(C', k)	$C' \in \{32, 64, 128\}$ $k \in \{3 \times 3, 5 \times 5\}$
Resblock	RB(C', k)	$C' \in \{32, 64, 128\}$ $k \in \{3 \times 3, 5 \times 5\}$
Max pooling	MP	–
Average pooling	AP	–
Summation	Sum	–
Concatenation	Concat	–

Because of the classification, the last output node is a softmax function (see p. 68) with the number of classes, and it is fully connected with all the input elements.

In Table 3.3, ResBlock is a description of a residual network shown in Fig. 3.15. ConvBlock is followed by a normal convolution operation, then batch normalization and a ReLU function (see p. 89). As a result of this operation, the feature map of input $M \times N \times C$ is converted to the feature map of output $M \times N \times C'$. Here, M, N, C, C' are the number of rows, columns, and input channels, respectively.

The concatenation function connects the two feature maps with different numbers of dimensions. That is, if two feature maps have a different number of rows or columns, the larger one is reduced by max pooling, and they are connected once they have the same dimensions. When the two input sizes are $M_1 \times N_1 \times C_1$ and $M_2 \times N_2 \times C_2$, the size of the feature map of the output of the concatenation operation is $\min(M_1, M_2) \times \min(N_1, N_2) \times (C_1 + C_2)$.

The summation function adds each element of the two feature maps. If the sizes of the two maps differ, down-sampling by max pooling is carried out. When the number of channels differs, the smaller one is added as an image of all zero values. That is, the feature map of the output is $\min(M_1, M_2) \times \min(N_1, N_2) \times \max(C_1, C_2)$.

Nagao et al. tested the effectiveness of this method using the CIFAR10 dataset (see p. 91). The mutation probability 0.05, number of columns (N_r) 5, number of rows (N_c) 30, and levels back (l) 10 were adopted as CGP parameters.

They experimented on the following two types as CGP function nodes:

- ConvSet: ConvBlock, Max pooling, Average pooling, Summation, Concatenation.
- ResBlock: Max pooling, Average pooling, Summation, Concatenation.

The maximum number of generations of CGP is 500 for ConvSet and 300 for ResSet.

In the first experiment, 45,000 images randomly selected from a total of 50,000 images were used as learning examples, and the remaining 5,000 images were used for the validation of CGP. Table 3.4 (left) shows the results for the full dataset. The proposed method is shown in the bottom two lines. For comparison, data from VGGNet [14] and ResNet [23], two frequently used deep learning methods, are

Table 3.4 Comparison of error rates from the CIFAR-10 dataset

Model	Full data		Small data	
	Error rate	#.Params ($\times 10^6$)	Error rate	#.Params ($\times 10^6$)
VGGNet [14]	7.94	15.2	24.11	15.2
ResNet [23]	6.61	1.7	24.10	1.7
CGP-CNN(ConvSet)	6.75	1.52	23.48	3.9
CGP-CNN(ResSet)	5.98	1.68	23.47	0.83

also shown. Figure 3.18 shows the structure of the network obtained as a result of evolution.

In the second experiment, 4,500 images randomly selected from a total of 5,000 images were used as learning examples, and the remaining 500 images were used for the validation. Table 3.4 (right) shows the results for the small dataset. Figure 3.19 shows the structure of the network obtained as a result of evolution.

The results of the experiment demonstrate that the method using CGP is equivalent to the conventional method in terms of classification accuracy. More important is the small number of parameters. In other words, the same results are obtained with a smaller network structure. This is meaningful in improving the disadvantages of conventional deep learning methods, namely the enormous learning time taken due to the large number of parameters and the considerably ad hoc design of the network structures.

3.2.3 *Differentiable Pattern-Producing Network*

Fernando et al. [1] proposed the differentiable pattern-producing network (DPPN) framework, which is an extension of the compositional pattern-producing network (CPPN, see Sect. 3.1.2) approach. DPPN uses the same node types as CPPN. There are two input nodes methods, i.e.,

- The same identity node as CPPN.
- A fully connected linear layer, mapping to a vector of the same dimension number.

At first, the DPPN is initialized with the structure shown in Fig. 3.20. This structure has two random hidden units.

The target task of DPPN is noise removal. Therefore, a denoising autoencoder is used. This is a framework for unsupervised learning. This network receives noisy data $\tilde{x}$ with respect to teaching data x and passes the data through the network layer as follows:

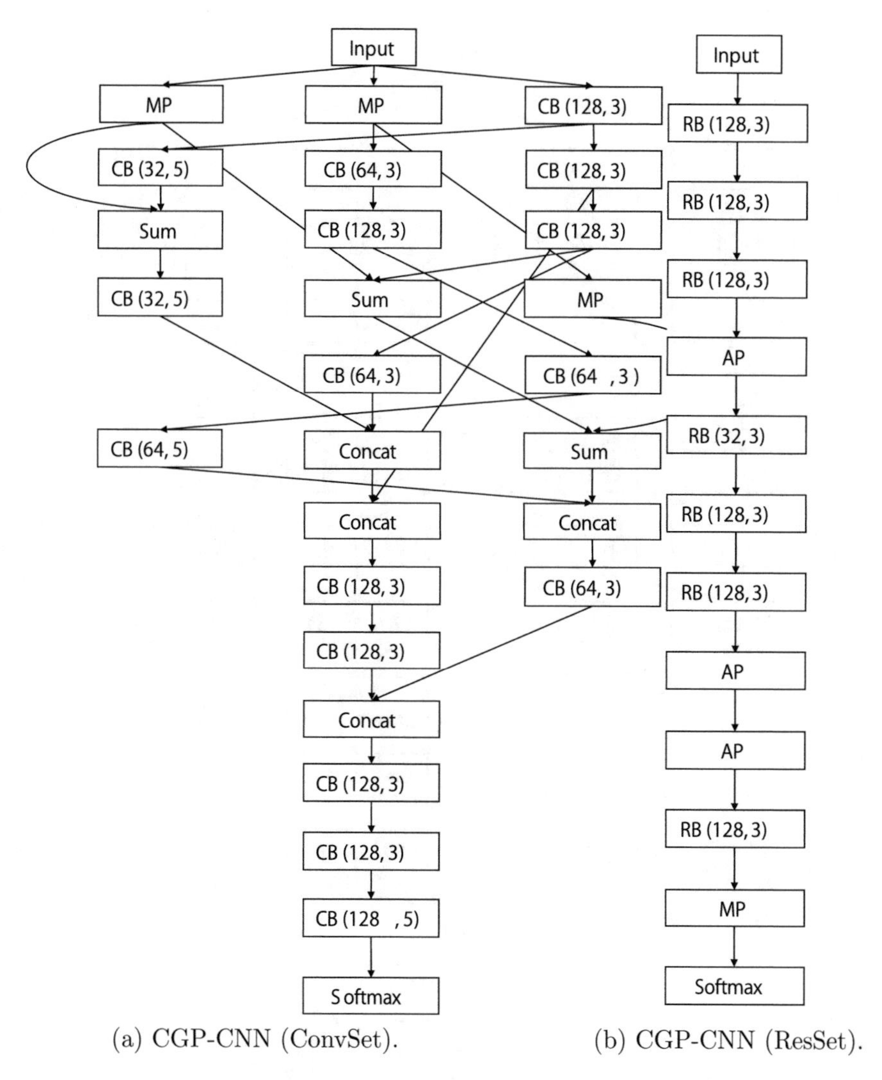

(a) CGP-CNN (ConvSet). (b) CGP-CNN (ResSet).

Fig. 3.18 Evolved convolutional neural network architectures (based on the full dataset) (Reproduced from [18])

$$f_\theta(\tilde{x}) = f^n_{\theta_n}(f^{n-1}_{\theta_{n-1}}(\cdots f^1_{\theta_1}(\tilde{x})\cdots)), \tag{3.1}$$

where $\theta = \{\theta_1, \ldots, \theta_n\}$ is a parameter. In this case, the mean square error (MSE) and binary cross-entropy (BCE[7]) of x and $\tilde{x}$ are calculated. In normal deep learning,

[7]BCE is a loss function $\mathcal{L}$ commonly used for a binary classification, which is a special case of multiclass cross-entropy. The definition is given as follows: $\mathcal{L}(\theta) =$

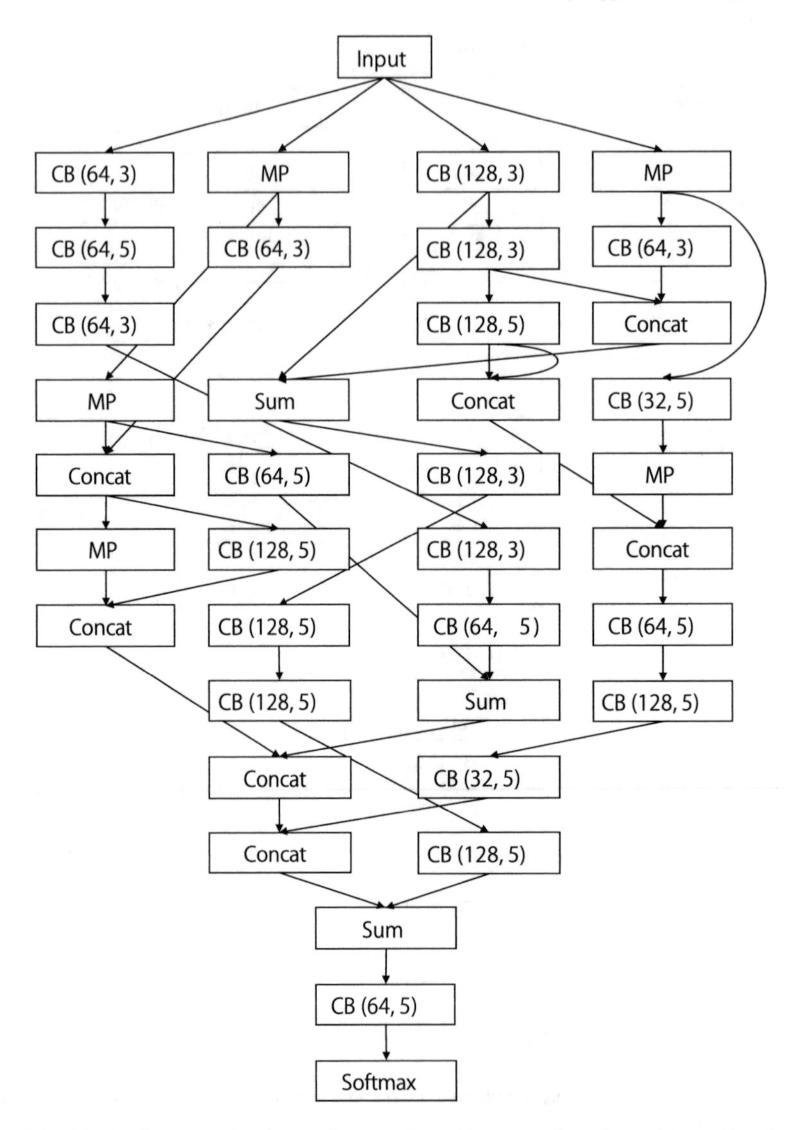

Fig. 3.19 Evolved convolutional neural network architectures (based on the smaller dataset) (Reproduced from [18])

errors are minimized by a gradient method. In DPPN, the output parameter p is mapped directly to parameter θ of the denoising autoencoder.

$-\frac{1}{n}\sum_{i=1}^{n}\left[y_i \log(p_i) + (1-y_i)\log(1-p_i)\right]$, where n is the number of samples, y_i is the sample label of ith sample, and p_i is the prediction for the ith sample. Smaller values indicate a better prediction.

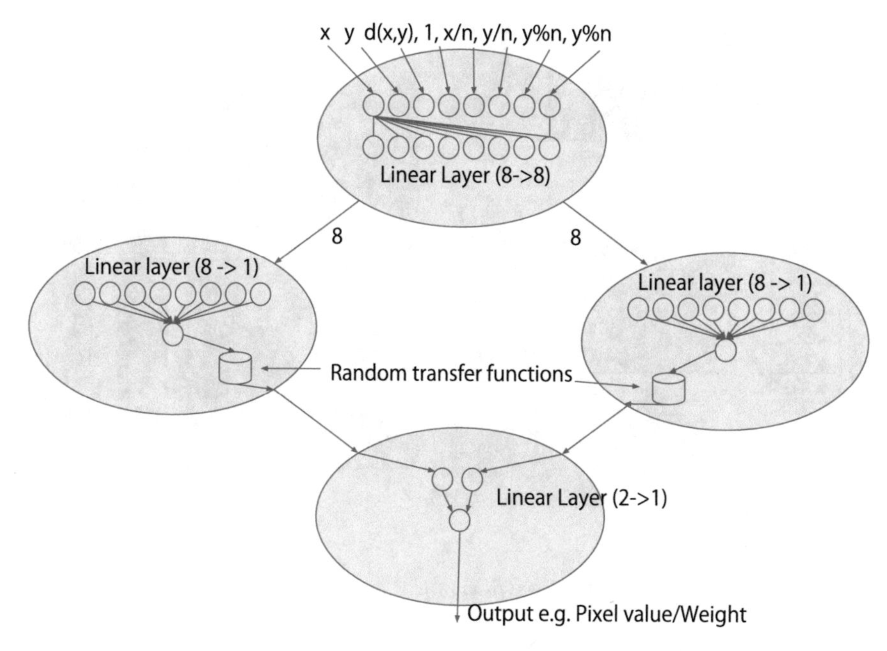

Fig. 3.20 Initial differential pattern-producing network topology (Reproduced from [1])

Essentially, evolution and learning algorithms are based on CPPN. However, Lamarckian evolution is also adopted and the learned (acquired) characteristics are inherited in DPPN. When calculating the fitness value, the results of the learning outcome from 1000 steps of DPPN according to minibatches of 32 sizes are used. The change of weight for the CPPN was backpropagated based on the slope of the loss function, but adaptive moment estimation (i.e., Adam [7]) is used in DPPN. This is a method of adaptively changing the learning rate using the gradient moment.

The target task is the reconstruction of MNIST data[8] with the 10% parameter set to 0. The overall flow of DPPN is shown in Algorithm 3.1 and Fig. 3.21. In order to evaluate the fitness value of each DPPN, the following process is repeated 1,000 times:

- A copy of the DPPN outputs to the autoencoder.
- A forward and a backward pass through the autoencoder with a MNIST minibatch.
- Backpropagation of these gradients through the DPPN.

DPPN indirectly learns parameters for a fully connected feed-forward denoising autoencoder. This network has the following features:

- one encoding layer with sigmoid activation functions.
- one decoding layer with sigmoid activation functions.
- a hidden layer consisting of 100 units (10×10 grid).

[8] http://yann.lecun.com/exdb/mnist/index.html.

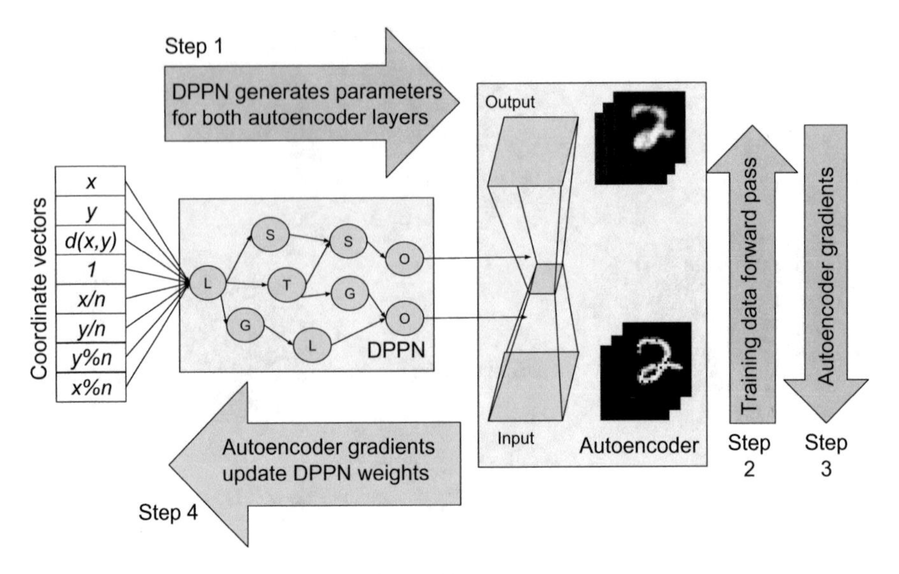

Fig. 3.21 DPPN algorithm (Reproduced from [1])

That is, the number of parameters (weight and bias) is $28 \times 28 \times 10 \times 10 \times 2 + 28 \times 28 + 10 \times 10 = 157,684$. The number of parameters is 1,000 times that of CNN for similar problems. The DPPN gives the output of two layers encoding and decoding these parameters. In order to obtain these parameters, 157,684 inputs were given to the DPPN. The input of the DPPN is the following input value of the autoencoder:

$$(x_{in}, y_{in}, x_{out}, y_{out}, D_{in}, D_{out}, \text{layer}, 1), \tag{3.2}$$

where x_{in} and y_{in} are the coordinates of the input neuron, and x_{out} and y_{out} are the coordinates of the output neuron. D_{in} and D_{out} are the distances from the centers of the input and output neurons. This DPPN generates $157,684 \times 2$ parameters. However, only the first $28 \times 28 \times 10 \times 10 + 10 \times 10$ elements of the first row and the second $28 \times 28 \times 10 \times 10 + 28 \times 28$ elements of the second row are used to encode the parameters of the autoencoder.

After learning, the negative BCE (or MSE) values for 100 random MNIST images are taken as the fitness of the DPPN. At the end of the run time, 1000 random MNIST images are tested and the BCE or MSE values are calculated.

Figure 3.22 shows the evolution process of reconstructing a handwritten image "2" by DPPN for a population size of 50. For the parameters, the crossover probability is 0.2, and the number of DPPN nodes is four for the initial individual. The right side of the figure illustrates Lamarckian evolution, in which the descendants inherit the learned weight. The middle figure shows Baldwinian evolution. Learning takes place, but the acquired characteristics are not inherited. Furthermore, the left figure shows Darwinian evolution, in which learning does not take place. Mutation

Algorithm 3.1 DPPN training algorithm

for 1,000 tournaments **do**
  Choose two DPPNs, $d_1, d_2 \in P$ ▷ P: population of DPPNs.
  $f_1 \leftarrow$ GetFitness(d_1)
  $f_2 \leftarrow$ GetFitness(d_2)
  Choose winner A and loser B from d_1 and d_2
  $B \leftarrow$ Mutate(Crossover(A, B))
end for

function GetFitness(DPPN d)
  for 1,000 steps **do**
    Parameters $\vec{p} \leftarrow d(\vec{c})$ ▷ $\vec{c}$: an input vector, Eq. (3.2).
    Copy $\vec{p}$ into denoising autoencoder parameters θ
    Choose minibatch x of MNIST images
    Generate noisy minibatch $\tilde{x}$
    Gradients $g_i \leftarrow \frac{\delta l(x, f_\theta(\tilde{x}))}{\delta w_i}$ w.r.t. DPPN weights ▷ l: a loss function, BCE or MSE.
    Follow Adam update to DPPN weights $\{w_i\}$ using $\{g_i\}$
  end for
  return fitness = $-MSE$ for 1,000 MNIST training images
end function

is added to the weight at a mutation probability of 0.001. Darwinian evolution is the closest of the three types of evolution to CPPN. In this example, the MSE values were 0.0036 (Lamarckian), 0.02 (Baldwinian), and 0.12 (Darwinian), respectively. Thus, Lamarckian evolution was found to perform the best. The number to recognize is inserted into the right figure (i.e., Target). Each figure shows the character generated in the evolution process of 1000 tournaments. The figure is sampled every 10 tournaments, and time progresses from the upper left to the lower right.

The experimental results show that DPPN is more efficient than CPPN. In addition, it was possible to evolve a simpler solution (i.e., network structure) than the conventional method of directly optimizing the weight of a large-scale network. For example, using DPPN to generate 157,684 parameters for a fully connected denoising autoencoder, a CNN structure with a fully connected feed-forward network embedded was acquired, and each hidden unit contained a blob-like 28×28 weight matrix. The blob moves smoothly in the receptive field of the hidden nodes. Furthermore, as a result of evolution, the network was also adapted to image cropping and magnification. For example, DPPN achieved 0.09 of the BCE value with MNIST test data with only 187 parameters. Even with data composed of Omniglot characters[9] or other datasets, better results were achieved than the network being directly encoded in the same way.

[9] 1623 different handwritten characters from 50 different alphabets. See https://github.com/brendenlake/omniglot for details.

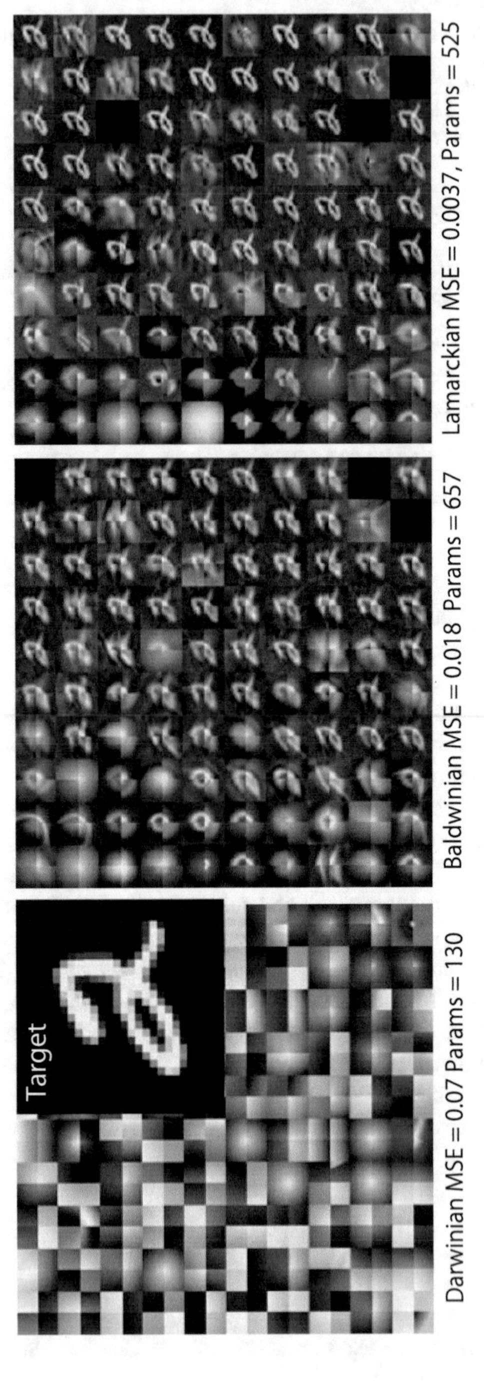

Fig. 3.22 Image reconstruction of a handwritten "2" (Reproduced from [1])

References

1. Fernando, C., Banarse, D., Reynolds, M., Besse, F., Pfau, D., Jaderberg, M., Lanctot, M., Wierstra, D.: Convolution by evolution–differentiable pattern producing networks. In: Proceedings of the Genetic and Evolutionary Computation Conference 2016. (GECCO16), pp. 109–116 (2016)
2. Hausknecht, M., Khandelwal, P., Miikkulainen, R., Stone, P.: HyperNEAT-GGP: A HyperNEAT-based Atari general game player. In: Proceedings of the Genetic and Evolutionary Computation Conference (GECCO 2012), pp. 217–224 (2012)
3. He, K., Zhang, X., Ren, S., Sun, J.: Deep residual learning for image recognition. In: Proceedings of the IEEE Conference on Computer Vision and Pattern Recognition (2016)
4. Huang, G., Liu, Z., Weinbergerz, K.: Densely connected convolutional networks. In: Proceedings of Computer Vision and Pattern Recognition (CVPR2017) (2017)
5. Iba, H.: Agent-Based Modeling and Simulation with Swarm. Chapman and Hall/CRC, London (2013)
6. Ioffe, S., Szegedy, C.: Batch normalization: accelerating deep network training by reducing internal covariate shift. In: Proceedings of the 32nd International Conference on Machine Learning, vol. 37, pp. 448–456 (2015)
7. Kingma, D.P., Ba, J.: Adam: a method for stochastic optimization. In: Proceedings of the 3rd International Conference on Learning Representations (ICLR2015)
8. Krizhevsky, A., Hinton, G.: Learning multiple layers of features from tiny images. Technical report 1. Computer Science Department, University of Toronto (2009)
9. Krizhevsky, A., Sutskerver, I. Hinton, G.E.: ImageNet classification with deep convolutional neural networks. In: Advances in Neural Information Processing Systems 25 (NIPS), pp. 1097–1105 (2012)
10. Miikkulainen, R., Liang, J., Meyerson, E., Rawal, A., Fink, D., Francon, O., Raju, B., Shahrzad, H., Navruzyan, A., Duffy, N., Hodjat, B.: Evolving deep neural networks (2017). arXiv:1703.00548
11. Netzer, Y., Wang, T., Coates, A., Bissacco, A., Wu, B., Ng, A.: Reading digits in natural images with unsupervised feature learning. In: Proceedings of NIPS Workshop on Deep Learning and Unsupervised Feature Learning (2011)
12. Rozenberg, G. (ed.): The Book of L. Springer, Berlin (1986)
13. Schrum, J., Miikkulainen, R.: Evolving multimodal behavior with modular neural networks in Ms. Pac-Man. In: Proceedings of the Genetic and Evolutionary Computation Conference (GECCO 2014), pp. 325–332 (2014)
14. Simonyan, K., Zisserman, A.: Very deep convolutional networks for large-scale image recognition. In: Proceedings of International Conference on Learning Representations (2014)
15. Stanley, K.O.: Compositional pattern producing networks: a novel abstraction of development. Genet. Program. Evolvable Mach. (Special Issue on Dev. Syst.) **8**(2), 131–162 (2007)
16. Stanley, K.O., Miikkulainen, R.: Evolving neural networks through augmenting topologies. Evol. Comput. **10**(2), 99–127 (2002)
17. Stanley, K.O., D'Ambrosio, D.B., Gauci, J.: A hypercube-based encoding for evolving large-scale neural networks. Artif. Life **15**(2), 185–212 (2009)
18. Suganuma, M., Shirakawa, S., Nagao, T.: A genetic programming approach to designing convolutional neural network architectures. In: Proceedings of the Genetic and Evolutionary Computation. Conference 2017 (GECCO2017), pp. 497–504 (2017)
19. Szegedy, C., Liu, W., Jia, Y., Sermanet, P., Reed, S., Anguelov, D., Erhan, D., Vanhoucke, V., Rabinovich, A.: Going deeper with convolutions. In: Proceedings of Computer Vision and Pattern Recognition (CVPR2016) (2016)
20. Unemi, T.: SBART2.4: Breeding 2D CG images and movies, and creating a type of collage. In: Proceedings of The Third International Conference on Knowledge-based Intelligent Information Engineering Systems, pp. 288–291 (1999)

21. Xie, L., Yuille, A.: Genetic CNN. In: Proceedings of IEEE International Conference on Computer Vision (ICCV)
22. Yao, X.: A review of evolutionary artificial neural networks. Int. J. Intell. Syst. **8**, 539–567 (1993)
23. Zagoruyko, S., Komodakis, N.: Wide residual networks (2016). arXiv: 1605.07146

Chapter 4
Machine Learning Approach to Evolutionary Computation

It is intriguing that computer scientists use the term genotype and phenotype when talking about their programs.
(John Maynard Smith, The Origins of Life: From the Birth of Life to the Origin of Language)

Abstract This chapter gives several methods of evolutionary computation enhanced with machine learning techniques. The employed machine learning schemes are bagging, boosting, Gröbner bases, relevance vector machine, affinity propagation, SVM, and k-nearest neighbors. These are applied to the extension of GP (Genetic Programming), DE (Differential Evolution), and PSO (Particle Swarm Optimization).

Keywords BagGP · BoostGP · Vanishing ideal GP (VIGP) · Kaizen programming · RVM-GP · Sequential sparse Bayesian learning algorithm · Particle swarm optimization based on affinity propagation (PSOAP) · ILSDE · SVC-DE TRAnsfer learning for DE (TRADE) · NENDE (k-NN classifier for DE)

4.1 BagGP and BoostGP

We divide the whole GP population into a set of subpopulations $\{P_1, P_2, \ldots, P_T\}$, in order to use the boosting and bagging techniques. When GP is applied to evolving individuals in each subpopulation P_t, the training set TR_t is sampled by means of the sampling method.

BagGP uses the bagging method, in which a training set of size N is sampled with replacement from the original instances for each subpopulation. Best individuals from each of the subpopulations vote in order to form a composite output for the testing data. The output value associated with most votes is considered as the final output. More precisely, we use the weighted median described below (**Step** 10) for this derivation.

H. Iba, *Evolutionary Approach to Machine Learning and Deep Neural Networks*, https://doi.org/10.1007/978-981-13-0200-8_4

BoostGP uses the following sampling method based on AdaBoost.R [12]. The following procedure (**Step** 2–**Step** 8) is repeated from the first subpopulation P_1 to the last subpopulation P_T.

Step1 Let t be set to 1. Each training pattern is assigned an equal weight, i.e., $w_i = 1$ for $i = 1, \ldots, N$, where N is the total number of the training data.

Step2 The probability that a training sample i is included in the training set TR_t is $p_i = w_i / \sum w_i$, where the summation is over all members of the training set. Pick N samples with replacement to form the training set.

Step3 Apply GP to the individuals in the subpopulation P_t with the above training set TR_t. The best evolved tree makes a hypothesis, i.e, $h_t : x \rightarrow y$.

Step4 Pass every member of the training set TR_t through this tree h_t to obtain a prediction $y_i^{(p)}(x_i)$ for $i = 1, \ldots, N$.

Step5 Calculate a loss $L_i = L(\mid y_i^{(p)}(x_i) - y_i \mid)$ for each training sample. The loss function L is defined later.

Step6 Calculate the averaged loss, i.e., $\overline{L} = \sum_{i=1}^{N} L_i p_i$.

Step7 Calculate the measure of confidence in the predictor, i.e., $\beta = \frac{\overline{L}}{1-\overline{L}}$.

Step8 Update the weights by using $w_i := w_i^{1-\overline{L_i}}$.

Step9 $t := t + 1$. If $t \leq T$, then go to **Step2**.

Step10 For a particular input x_i in the test data, each of the T acquired trees, i.e., the best evolved individuals from the subpopulations, makes a prediction h_t for $t = 1, \ldots, T$. Obtain the cumulative prediction h_f using the T predictors:

$$h_f = \min \left\{ y \in Y : \sum_{t:h_t \leq y} \log(1/\beta_t) \geq \sum_t \log(1/\beta_t) \right\}. \quad (4.1)$$

In **Step5**, the loss function L may be of any form as long as $L \in [0, 1]$. We use the following loss function:

$$L_i = \frac{\mid y_i^{(p)}(x_i) - y_i \mid}{\max_{i=1,\ldots,N} \mid y_i^{(p)}(x_i) - y_i \mid}. \quad (4.2)$$

β is a measure of confidence in the predictor, i.e., the lower the β value is, the higher the confidence. In **Step9**, the smaller the loss is, the more the weight is reduced, which makes the probability smaller that this pattern will be chosen as a member of the training set for the next subpopulation. **Step10** represents the weighted median. Suppose that each best tree h_t has a prediction $y_i^{(t)}$ on the ith pattern and an associated β_t value and that the predictions are relabeled for pattern i as follows:

$$y_i^{(1)} < y_i^{(2)} < \cdots < y_i^{(T)}. \quad (4.3)$$

Then sum the $\log(1/\beta^t)$ until the smallest t is reached so that the inequality is satisfied. The prediction from that best tree t is taken to be the final prediction. If all β_t's are equal, it is the same as the median calculation.

We study the performance of BagGP and BoostGP empirically. Suppose that the total population size is set to be N_{psize}. This is divided into 10 subpopulations of size $N_{psize}/10$ for running BagGP and BoostGP. For the sake of comparison, we also experiment in running the canonical GP, i.e., a standard GP without any sampling method for the single population of size N_{psize}. The parameters used for the experiments are shown in Table 4.1.

GP is applied to the following symbolic regression problems: The goal of symbolic regression is to estimate a generative function for some given data, and symbolic regression appears in a wide range of research fields (see Sect. 4.2.1 for detailed discussion).

Experiment 1 Discovery of trigonometric identities.
Koza used GP to find a new mathematical expression, in symbolic form, that equals a given mathematical expression, for all values of its independent variables [Koza92, chap. 10.1]. In the following experiment, our goal is to discover trigonometric identities, such as

$$\cos 2x = 1 - \sin^2 x. \tag{4.4}$$

The training data consist of 100 pairs, i.e., $\{(x, \cos 2x + \Re)\}$, in which a random noise $\Re$ between –0.01 and 0.01 is added to the precise value. The testing data consist of 100 input–output pairs, i.e., $\{(x, \cos 2x)\}$, where no random noise is added to the output value.

Experiment 2 Predicting a chaotic time series.
The Mackey–Glass time series is generated by integrating the following delay differential equation and is used as a standard benchmark for prediction algorithms:

$$\frac{dx(t)}{dt} = \frac{ax(t-\tau)}{1 + x^{10}(t-\tau)} - bx(t), \tag{4.5}$$

with $a = 0.2$, $b = 0.1$, and $\tau = 17$. The trajectory is chaotic and lies on an approximately 2.1-dimensional strange attractor.

For the sake of comparison, all the parameters chosen were the same as those used in the previous study [Oakley94, p. 380, Table 17.3], except that the terminal set consisted of ten past data for the short-term prediction (see Table 4.1). We use the first 100 time sequences for the training data and the next 400 time sequences for the testing data.

Experiment 3 Boolean concept formation (6-multiplexer).
To show the effectiveness as a Boolean concept learner, we conducted a simple experiment (6-multiplexer), in which the goal function is the multiplexer function of 6 variables:

Table 4.1 GP parameters

Parameters of GP	Problem name		
	cos(2x) (Exp.1)	Chaos (Exp.2)	6-multiplexer (Exp.3)
Terminal symbols	{X,1.0}	$\{X(t-1), X(t-2), \ldots, X(t-10), \Re\}$	$\{a_0, a_1, d_0, d_1, d_2, d_3\}$
Non-terminal symbols	$\{+, -, *, \%, \sin\}$	$\{+, -, \times, \%, \sin, \cos, \exp_{10}\}$	{AND, OR, NAND, NOR}
No. of subpopulations	10	10	10
subpopulation_size	200	1024	1024
max_depth_for_new_trees	6	6	6
max_depth_after_crossover	17	17	17
max_mutant_depth	4	4	4
grow_method	GROW	GROW	GROW
selection_method	TOURNAMENT	TOURNAMENT	TOURNAMENT
tournament_K	6	6	6
crossover_func_pt_fraction	0.1	0.1	0.1
crossover_any_pt_fraction	0.7	0.7	0.7
fitness_prop_repro_fraction	0.1	0.1	0.1

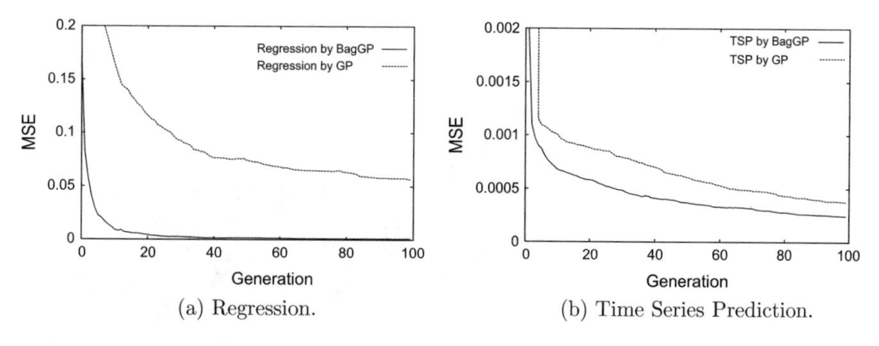

(a) Regression. (b) Time Series Prediction.

Fig. 4.1 Experimental results (Generation vs. MSE)

Table 4.2 Summary of experimental results

Problem name	cos(2*x*) (Exp.1) MSE	Chaos (Exp.2) MSE	6-multiplexer (Exp.3) Success Gen.
GP	0.056606	0.000375	5.15
BagGP	0.001051	0.000244	4.80
BoostGP	0.019296	0.000237	4.88

$$f(a_0, a_1, d_0, d_1, d_2, d_3) = \overline{a}_0\overline{a}_1 d_0 \vee a_0\overline{a}_1 d_1 \vee \overline{a}_0 a_1 d_2 \vee a_0 a_1 d_3 \tag{4.6}$$

The inputs to the Boolean 6-multiplexer function consist of 2 address bits a_i and 2^2 data bits d_i. The output is the Boolean value of the particular data bit that is singled out by the address bits of the multiplexer. For example, if $a_0 = 1$ and $a_1 = 0$, then the multiplexer singles out the data bit d_2 because $10_2 = 2$.

For both the training and testing datasets, we use the whole set of 64 pairs of 6 binary input variables, i.e., $\{(a_0, a_1, d_0, d_1, d_2, d_3) \in \{0, 1\}^6\}$ and their output values derived by Eq. (4.6).

Figure 4.1 shows the experimental results, which plots the mean square errors (MSE) for the testing data with generations, averaged over 20 runs. Figure 4.2 plots the standard fitness values of BoostGP with generations. As shown in the figure, the BoostGP was able to track the oscillation and adapt to a new training data quickly. Table 4.2 summarizes the experimental results. The table compares the MSE values for the test data at the final generation for Experiments 1 and 2. For Experiment 3, the success generations to obtain the solution tree are listed. All these values are averaged over 20 runs. This comparison clearly shows the superiority of BagGP and BoostGP over the standard GP. The robustness of evolved trees by BagGP and BoostGP was enhanced in the sense that the validate fitness values, i.e., the mean square errors for the testing data, were reduced. The performance difference between BagGP and BoostGP was not statistically significant.

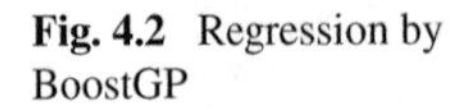

Fig. 4.2 Regression by BoostGP

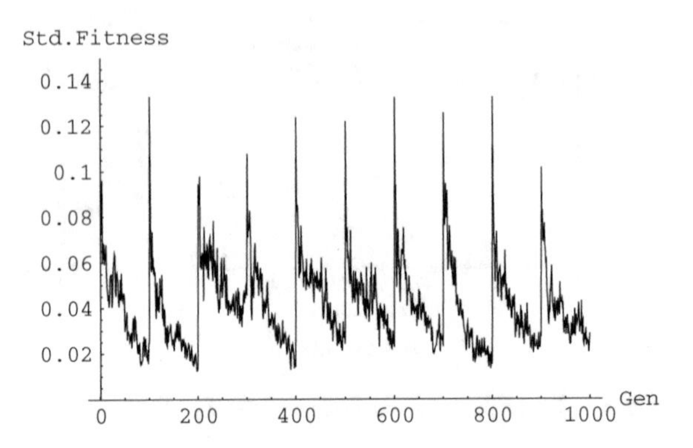

BagGP and BoostGP have been successfully applied to real-world tasks, i.e., financial data prediction. The performance is discussed by the comparison with the traditional GP in view of the bloat effect [17].

4.2 Vanishing Ideal GP: Algebraic Approach to GP

This section describes a novel GP method (VIGP: Vanishing Ideal GP [21]), in which a reduction phase by means of ideal bases is introduced just prior to the evaluation phase in GP. VIGP can be applied to symbolic regression of rational polynomials, which have considerable expressive power. Hence, VIGP is able to achieve a good fitting for a wide range of function classes.

4.2.1 Symbolic Regression and GP

GP provides a powerful approach to symbolic regression in that it does not require models of functions to be fixed. However, it is known that GP suffers from a phenomenon known as bloat, meaning that candidate functions attain an excessively complicated form during the search, which is undesirable in many applications. While the majority of approaches for regulating bloat introduce anti-bloat genetic operators or anti-bloat selection schemes, most of these are derived from heuristics and/or require well-tuned hyperparameters. In this section, we describe a novel approach in which genetic trees of GP are reduced during the search using a basis of a set of polynomials (*vanishing ideal*) that are equivalent to zero for the data points of symbolic regression.

In GP, functions expressed in a tree form are randomly generated, mated, and mutated during the search. Positive bloat is a well-known phenomenon in which the trees in GP become rapidly complicated during the search, resulting in the search

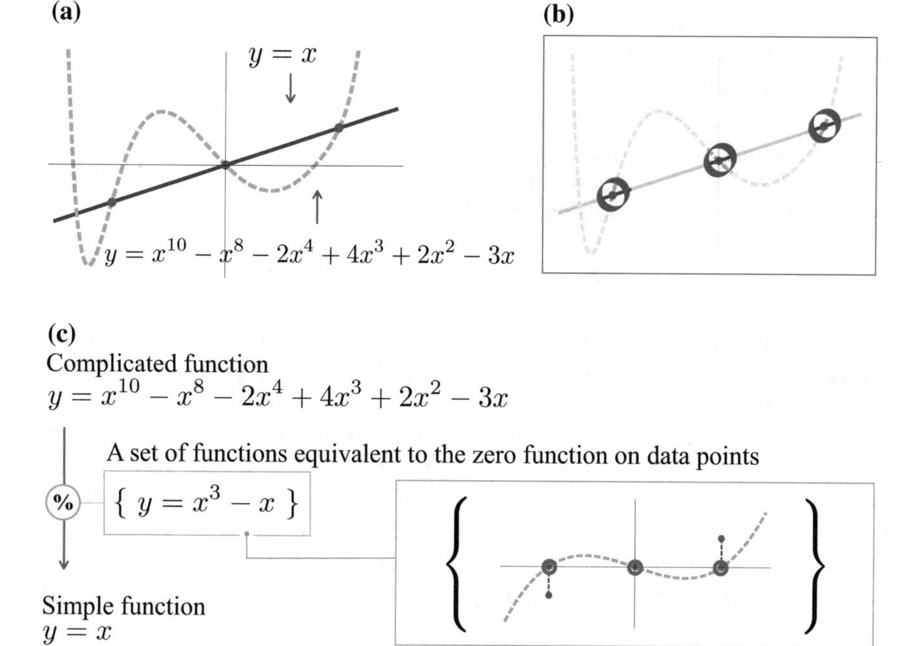

Fig. 4.3 Vanishing ideal: **a** Two solutions for a symbolic regression problem with data consisting of the red points. One is a simple solution (solid line; blue), while the other is more complicated (dashed line; green). **b** From the point of view of GP, these two functions are equivalent because the behavior of them for the data points is the same. **c** The key idea of VIGP. A complicated function can be reduced to a simple function by dividing it by a finite set of functions that are equivalent to the zero function for the data points (dashed line; cyan). In general, the set contains more than one element, while it only has one element in the two-dimensional case shown in the example

having a low efficiency and solutions being excessively complicated [28]. One reason for this is that the fitness of a tree in GP is based on its mean squared error for the data points. In other words, the fitness is based only on the behavior of functions for the data points. As a consequence, when a simple function (e.g., Fig. 4.3a; solid line) and complicated function (e.g., Fig. 4.3a; dashed line) are given, these are viewed as exactly the same by GP as long as they take the same values for the data points (Fig. 4.3b).

Focusing on the behavior of functions for data points, we notice that there are also some functions that are equivalent to zero for the data points (vanishing functions) while not equivalent to zero in an algebraic sense. The key idea of VIGP is that simpler functions can be obtained from complicated functions by exploiting such vanishing functions. Now, the main question is how to find such vanishing functions that they most effectively reduce a complicated function. This is a very difficult problem in general, because there are infinitely many vanishing polynomials. However, when the functions are restricted to polynomials, it is known that a set of vanishing polynomials for data points (vanishing ideal [26]), which also has an infinite number of elements, can be described by a finite set of vanishing polynomials, or a basis of a vanishing ideal [5]. By exploiting this mathematical result, we reduce complicated functions to simple functions by dividing them by a basis of a vanishing ideal (Fig. 4.3c).

Suppose that we have a list $\mathcal{X} = \{\boldsymbol{x}_1, \ldots, \boldsymbol{x}_N\}$ of N data points in n dimensions and a list $\mathcal{T} = \{t_1, \ldots, t_N\}$ of target values at each point in $\mathcal{X}$. Our goal is to find an n-variate polynomial f satisfying the following condition:

$$t_i = f(\boldsymbol{x}_i), \quad \forall i \in \{1, \ldots, N\}. \tag{4.7}$$

That is, to find an n-variate polynomial f that returns the target values at each point. For the sake of simplicity of notation, we use $f(\mathcal{X})$ to denote $(f(\boldsymbol{x}_1), \ldots, f(\boldsymbol{x}_N))$ in some cases.

One common problem in symbolic regression is that there are an infinite number of solutions that satisfy the condition in Eq. (4.7) for a given dataset $\{\mathcal{X}, \mathcal{T}\}$. Note that this argument holds for general functions, while in VIGP we restrict f to being a polynomial. Now, suppose that we have a function f satisfying the condition Eq. (4.7). Let us consider a function g that vanishes on the data points $\mathcal{X}$, i.e., $g(\mathcal{X}) = \mathbf{0}$. In this case, the following function $\tilde{f}$ also satisfies Eq. (4.7):

$$\tilde{f} = f + hg, \tag{4.8}$$

where h is a function that does not attain infinity for the data points. This argument can be confirmed as follows:

$$\begin{aligned} \tilde{f}(\mathcal{X}) &= f(\mathcal{X}) + h(\mathcal{X}) \circ g(\mathcal{X}), \\ &= f(\mathcal{X}) + h(\mathcal{X}) \circ \mathbf{0}, \\ &= f(\mathcal{X}), \end{aligned}$$

where $\circ$ is an element-wise product of a pair of vectors. Because there are an infinite number of such h, there also exist infinitely many $\tilde{f}$. In most applications, the *simplest* solution is preferred, while the kind of form that is considered simple depends on the application (e.g., low degree forms or those consisting of few terms). As already mentioned, while GP has the advantage of flexibility, in that it can estimate the model of functions, it suffers from bloat, which results in excessively complicated and thus undesirable functions. In this section, we describe an approach to finding simple solutions, in that they do not contain any vanishing functions as the second term in Eq. (4.8).

4.2.2 *Vanishing Ideal*

Let $\mathcal{P}$ be a set of n-variate polynomials. A vanishing ideal $I_{\mathcal{X}}$ of data points $\mathcal{X}$ is a set of polynomials that vanish on $\mathcal{X}$:

$$I_{\mathcal{X}} = \{g \in \mathcal{P} \mid g(\mathcal{X}) = \mathbf{0}\}.$$

According to the Hilbert basis theorem, $I_{\mathcal{X}}$ can be spanned by a finite set of polynomials [5]. That is, when $I_{\mathcal{X}}$ is spanned by a set of s polynomials $\mathcal{G} = \{g_1, \ldots, g_s\}$ and it holds that

$$\forall g \in I_{\mathcal{X}}, \quad g = \sum_{i=1}^{s} h_i g_i, \tag{4.9}$$

where $h_i \in \mathcal{P}$, then $\mathcal{G}$ is called a basis of the vanishing ideal $I_{\mathcal{X}}$. While Eq. (4.9) is defined for polynomials in $I_{\mathcal{X}}$, the following holds for any polynomial $f \in \mathcal{P}$:

$$f = \sum_{i=1}^{s} h_i g_i + r, \quad \text{s.t. } \mathrm{LM}(f) \geq \mathrm{LM}(r), \tag{4.10}$$

where $h_i, r \in \mathcal{P}$ and $\mathrm{LM}(\cdot)$ is the leading monomial of a polynomial with respect to the purely lexicographical ordering (i.e., $<_{PL}$, see p. 49 or the details of this explanation). r is always less than or equal to f in Eq. (4.10) in a monomial ordering sense.

Note that f and r in Eq. (4.10) are different in general, but they are equivalent to each other on the data points $\mathcal{X}$:

$$\begin{aligned} f(\mathcal{X}) &= \sum_{i=1}^{s} h_i(\mathcal{X}) \circ g_i(\mathcal{X}) + r(\mathcal{X}), \\ &= \sum_{i=1}^{s} h_i(\mathcal{X}) \circ \mathbf{0} + r(\mathcal{X}), \\ &= r(\mathcal{X}). \end{aligned}$$

Therefore, Eq. (4.10) can be interpreted as a reduction of f to r, while preserving exactly the same behavior for the data points. From the point of view of GP, these have exactly the same fitness when the fitness function is defined as the mean square error from the target values. As already mentioned, simpler functions are preferred in most applications. Therefore, we assume that r is better than f, and introduce an algebraic reduction method instead of adding a penalty term into the fitness function, as is commonly done.

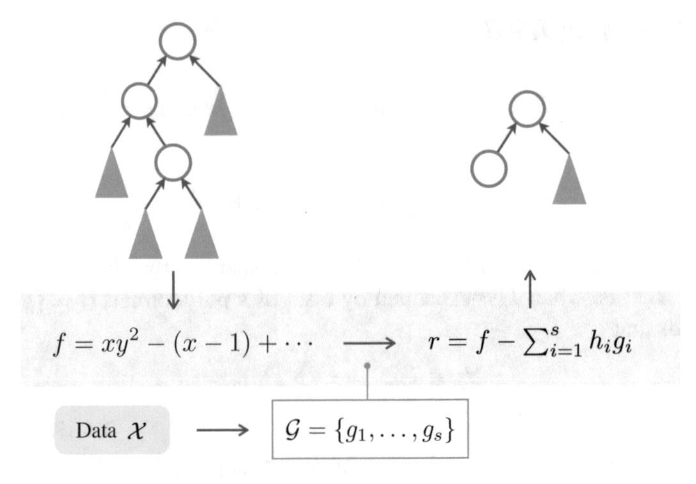

Fig. 4.4 The pipeline for removing vanishing polynomials. A tree is converted to a polynomial, then reduced using a basis of the vanishing ideal of the set of data points, which is computed beforehand, and finally is reconverted to a tree

4.2.3 VIGP: Reduction Process

In the framework of VIGP, a reduction phase based on Eq. (4.10) is introduced just prior to the evaluation phase in GP. Here, we explain this reduction step. Suppose that we have a tree T_f of a polynomial f. Then, we reduce the tree according to the following procedure:

1. Convert the tree T_f to a polynomial f, and expand it. The semantic introns[1] are removed in this step.
2. Compute the residual polynomial r of f, with respect to the vanishing ideal $I_{\mathcal{X}}$ of the set of data points $\mathcal{X}$, as Eq. (4.10).
3. Construct a tree T_r of the residual polynomial r.

We provide the pipeline for this procedure in Fig. 4.4.

Next, we provide details of the second step of the procedure. In this step, the following questions arise:

- How to obtain a basis $\mathcal{G} = \{g_1, \ldots, g_s\}$ of the vanishing ideal $I_{\mathcal{X}}$?
- How to reduce a polynomial f to a residual polynomial r using a basis $\mathcal{G}$ as Eq. (4.10)?

As for the first point, we use the BM algorithm for symbolic regression with discrete-valued data points and use the ABM algorithm for real-valued data points (see

[1]Introns are structures that do not affect fitness values. In GP, two types of introns are known: (1) semantic introns: code segments that are executed but have no effect on the overall result, e.g., (+ 3 (− x x)), (2) syntactic introns: non-executed code segments, e.g., (and false (+ 2 3)).

Sect. 2.2.7 and p. 51 for details). Note that it is only necessary to compute the basis $\mathcal{G}$ once, because it only depends on the list of data points $\mathcal{X}$.

After obtaining $\mathcal{G}$, we reduce a polynomial f to a residual polynomial r using a basis $\mathcal{G}$, as Eq. (4.10), which has been described in p. 51. Remember that, with the Gröbner base or the border basis, f can be reduced to r in a unique form. Thus, one can remove duplicate polynomials from a population of GP.

Following the reduction, the polynomial is converted to a tree. At this point, we first reconstruct trees from each monomial and its coefficients. If there are m monomials in a polynomial, then m trees are reconstructed. These are then concatenated with addition operators in the order of high to low.

While the reduction and reconstruction may destroy the structures obtained during the search in GP, their execution is worthwhile when both semantic introns and vanishing polynomials can be removed. Some trees only contain semantic introns, but no vanishing polynomials. In this case, it may be favorable not to perform the reduction, but to preserve the good tree structures as far as possible. This reduction rule is an option for VIGP, and in VIGP, we approximately follow it by exploiting the fact that the degree of a polynomial f can be bounded by the height of its tree T_f:

$$deg(f) \leq 2^{H(T_f)},$$

where $deg(\cdot)$ is the total degree of a polynomial and $H(\cdot)$ is the height of a tree. With this bound and the minimum total degree $deg_{\min}(\mathcal{G})$ of the polynomials in $\mathcal{G}$, we only apply our reduction to trees satisfying the following condition:

$$\lceil \log_2 (deg_{\min}(\mathcal{G})) \rceil \leq H(T_f),$$

where T_f is the tree of a polynomial f and $\lceil \cdot \rceil$ is the ceiling function, which maps a real number to the smallest following integer. This reduction rule preserves good tree structures obtained during the search in GP, while it retains some semantic introns. There is an only small difference between VIGP with and without this option in the results of our experiments. We introduce it mainly to increase the computation speed.

4.2.4 VIGP Versus GP Comparison

In this section, we compare VIGP with standard GP and demonstrate that VIGP successfully regulates bloat and achieves an efficient search.

First, we show that bloat easily occurs in GP. In Fig. 4.5, the average height of trees (solid line; red) and the minimum fitness in the population (dashed line; blue) are plotted with various crossover probabilities. Note that here the fitness is defined as the mean square error (MSE), so that trees with a smaller fitness are preferred to those with a higher fitness. The crossover probability varies from 0.1 to 0.9 in intervals of 0.1, and the mutation probability was selected such that the sum of

Table 4.3 Experimental conditions

	Figs. 4.5/4.6/4.7/4.9
Population size	500/500/500/200
The number of generations	200/200/200/100
Crossover probability	– /0.9/0.9/0.9
Mutation probability	– /0.1/0.1/0.1
Elite rate	0.01/0.01/0.01/0.01

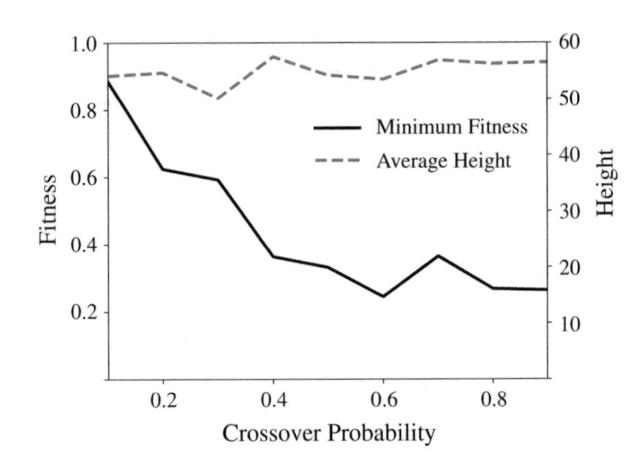

Fig. 4.5 The average height (solid line) and minimum fitness (dashed line) of trees in the final population in GP. A smaller fitness is better. The crossover probability varies from 0.1 to 0.9, in intervals of 0.1. At each crossover probability, the average height is extremely large. That is, bloat occurs. The value given at each point is the average of fifty trials

the two probabilities is equal to one. Additional experimental conditions were as described in Table 4.3. We employed the following function as the target function:

$$f_1^\star = x^4 - x^3 + \frac{y^2}{2} - y, \tag{4.11}$$

which is a benchmark function (Keijzer-12, [24]) for symbolic regression. Sixteen data points were sampled from $-1.5 \leq x, y \leq 1.5$, at even intervals. The scores at each crossover probability are given as the average of fifty trials. As shown in Fig. 4.5, the average height of trees in a population is very large regardless of the crossover probabilities. In other words, bloat occurs. The number of nodes increases exponentially with the height of trees, and thus, memory resources are heavily exhausted by bloat. In addition, the exponential increase in function nodes implies an exponential increase in operations when a tree is evaluated, which results in a long evaluation time.

We compare the performance of VIGP with that of GP for symbolic regression with real-valued data. For N data points $\mathcal{X}$ and their target values $\mathcal{T}$, the fitness of a tree $T_{\hat{f}}$ is here defined as follows:

$$\text{fitness} = \frac{1}{N}\sum_{i=1}^{N}\left(\hat{f}(\boldsymbol{x}_i) - t_i\right)^2 + \lambda H(T_{\hat{f}}), \tag{4.12}$$

where $\boldsymbol{x}_i \in \mathcal{X}$, $t_i \in \mathcal{T}$, and λ is a hyperparameter to control the penalty weight for the height of trees. Again, note that trees with a lower fitness value are preferred here. We compared VIGP to GP with $\lambda = 0, 10^{-4}, 0.01, 0.1, 1$. A small value $\lambda = 10^{-4}$ was also given for VIGP, so as to select the simpler solution of two trees with exactly the same MSE. The experimental conditions are summarized in Table. 4.3. For the target functions, we adopted $f_1^\star$ from Eq. (4.11) and the function

$$f_2^\star = \frac{x^3}{5} + \frac{y^3}{2} - y - x,$$

which is also a benchmark function (Keijzer-15, [24]) for symbolic regression. Sixteen data points were sampled from $-1.5 \leq x, y \leq 1.5$ at even intervals.

The results are shown in Fig. 4.6. The average height and the MSE given in Fig. 4.6 represent the average of twenty trials. From Fig. 4.6a, c, one can see that VIGP

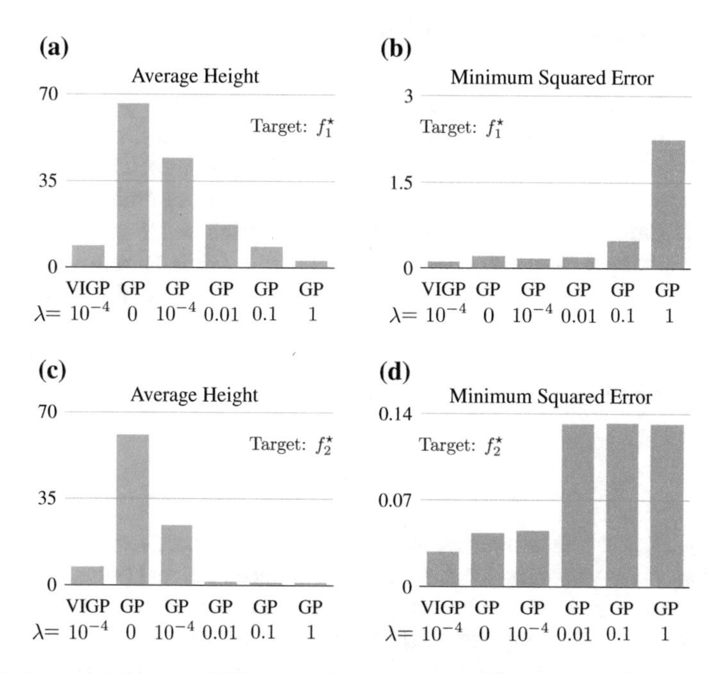

Fig. 4.6 Comparison between VIGP and GP for real-valued symbolic regression. **a**, **c** Comparison with respect to the average height of trees of the final generation, where the target function is $f_1^\star$ in **a** and $f_2^\star$ in **c**. VIGP achieved a moderate average height, while GP with a small value of λ resulted in positive bloat and GP with a large value of λ resulted in negative bloat. **b**, **d** Comparison with respect to the minimum MSE in the final generation, where the target function is $f_1^\star$ in **b** and $f_2^\star$ in **d**. VIGP achieved the smallest MSE

successfully regulated the height of trees, while GP with a small value of λ fails to do so. As the value of λ increases, GP also achieved a shorter average height of trees. However, as can be seen from Fig. 4.6b, d, while VIGP found a good solution with small a MSE, GP with a large value of λ only found poor solutions. By comparing these results, one can also observe that without carefully selecting the value of λ for each problem, the performance of GP deteriorates, with an extremely short average height, known as negative bloat (e.g., GP with a large value of λ in Fig. 4.6a, c), or a large MSE (e.g., GP with a large value of λ in Fig. 4.6b, d). Again, note that this problem does not occur in VIGP, because it involves no hyperparameters. The basis of the vanishing ideal is computed only from data points, and when the data points imply that complicated polynomials are required for a good fitting, a basis is obtained that leaves rather complicated polynomials following reduction.

Here, we measure the MSE using not unseen test data but training data. Using unseen test data is usually introduced to measure whether estimated solutions avoid overfitting. In VIGP, we are focusing on regulating bloat, and overfitting is another independent phenomenon as pointed out in [6, 34].

4.2.5 VIGP for Rational Polynomials

In the previous section, we used addition and multiplication as operators. Here, we introduce a division operator in order to extend the function class to rational polynomials. Rational polynomials are known as a class of functions that have considerable expressive power. While the Taylor expansion approximates arbitrary functions in polynomial form, the Padé approximation gives a much better approximation in a rational polynomial form [1]. Therefore, although VIGP is currently restricted to rational polynomials, we consider that the expressive power is sufficient for practical applications.

In the reduction phase of VIGP, a tree T_f is converted to a rational polynomial of the following form:

$$f = \frac{f_{\text{num}}}{f_{\text{den}}}, \tag{4.13}$$

where $f_{\text{num}}, f_{\text{den}} \in \mathcal{P}$. The numerator and denominator, respectively, are then reduced using a basis of a vanishing ideal. Finally, the reduced rational polynomial is converted to a tree by concatenating trees of the numerator and denominator with a division operator. We compared VIGP with GP in the case of rational polynomials, and the results are shown in Fig. 4.7. The average height and minimum MSE in the final generations are given as the average of twenty trials. The experimental conditions are described in Table 4.3, and the following target functions were used:

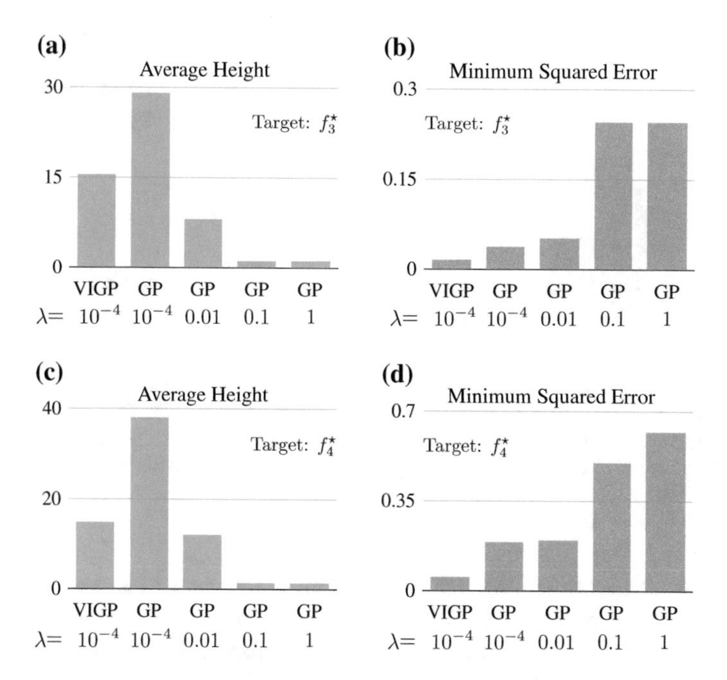

Fig. 4.7 Comparison between VIGP and GP for real-valued symbolic regression of rational polynomials. **a, c** Comparison with respect to the average height of trees of the final generation, where the target function is $f_3^\star$ in **a** and $f_4^\star$ in **c**. VIGP achieved a moderate average height, while GP with a small value of λ resulted in positive bloat and GP with a large value of λ resulted in negative bloat. **b, d** Comparison with respective to the minimum MSE in the final generation, where the target function is $f_3^\star$ in **b** and $f_4^\star$ in **d**. VIGP achieved the smallest MSE

$$f_3^\star = \frac{8}{2 + x^2 + y^2},$$
$$f_4^\star = \frac{1}{1 + x^{-4}} + \frac{1}{1 + y^{-4}},$$

which are benchmark functions from [24] to [27], respectively. Twenty-five data points were sampled from $-2 \leq x, y \leq 2$ at even intervals. The fitness of trees was defined as Eq. (4.12). As can be seen from Fig. 4.7a, when the target function was $f_3^\star$, VIGP achieved a moderate average height for the population in the final generation, while GP with inadequate values of λ resulted in a positive or negative bloat. At the same time, VIGP achieved the smallest MSE in the final generation, and GP with inadequate values of λ only achieved a rather large MSE (Fig. 4.7b). A similar trend can be observed when the target function is $f_4^\star$ (Fig. 4.7c, d). VIGP achieved a moderate average height and the smallest MSE, while GP with values of λ that were too small or large achieved significantly poorer results. The results of GP with $\lambda = 0$ are omitted, because in that case trees were affected by an extreme positive bloat, and the computation had still not stopped after a few days.

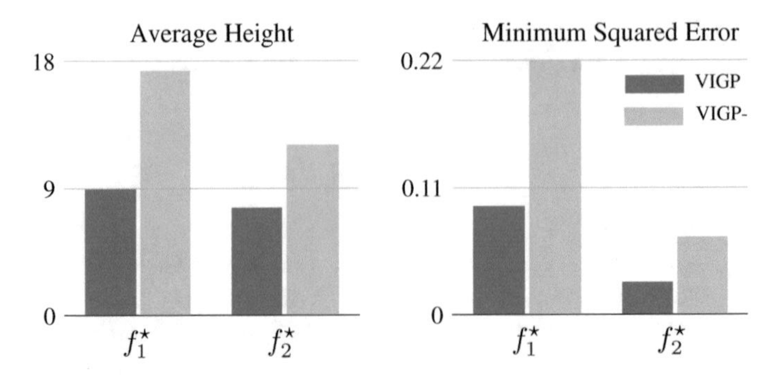

Fig. 4.8 Comparison between regular VIGP and VIGP where only semantic introns are removed (VIGP$^-$). **a** Comparison with respect to the average height of trees of the final generation. VIGP successfully regulated the average height of trees in both target functions. **b** Comparison with respective to the minimum MSE in the final generation. VIGP achieved the smallest MSE for both target functions

We have demonstrated that VIGP successfully regulates bloat and finds better solutions in terms of the MSE. However, in VIGP not only vanishing polynomials are removed, but also the semantic introns. Therefore, it is possible that it is not the removal of vanishing polynomials, but that of the semantic introns, that gives the major contribution to the results. We compare VIGP to VIGP without removing vanishing polynomials. We call the latter VIGP$^-$. In VIGP$^-$, trees were converted into polynomials, expanded, and reconverted to trees as in ordinary VIGP, and the semantic introns were removed during this procedure. We compared them in the case of real-valued symbolic regression described in the previous section. All of the experimental conditions are the same as for that experiment. As can be seen from Fig. 4.8, VIGP achieved a shorter average height and smaller MSE for both target functions. The differences are sufficiently large so that we can conclude that the removal of vanishing polynomials has a non-negligible effect in real-valued symbolic regression.

4.2.6 *VIGP for the 6-Parity Problem*

Now, we apply VIGP to the 6-parity problem [28], the goal of which is to construct a logic function that returns one only in the case that an even number of the six inputs are one. In general, logic operators (e.g., AND, OR) are used for function nodes of GP's trees. The logic functions can be mapped to the Boolean polynomials, that is, to polynomials whose coefficients are zero or one where the calculation is performed under modulus two. In general, there is no one-to-one mapping between a logic function and Boolean polynomial. In such case, which mapping to select is not trivial, while the selection largely affects the height of GP tree. In order to avoid this

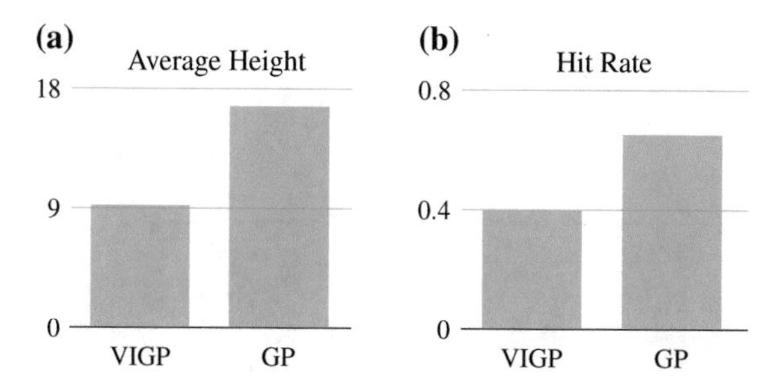

Fig. 4.9 Comparison between VIGP and GP in the 6-parity problem. **a** Comparison with respect to the average height of trees of the final generation. VIGP successfully regulated the average height of trees. **b** Comparison with respective to the hits rate. VIGP resulted in a lower hit rate than that of GP

effect, we here only use the AND operator and XOR operator, which respectively correspond to multiplication and addition in the Boolean domain.

For comparison, we defined the hit rate as the ratio between the number of trials in which the solutions of the 6-parity problem are found and the total number of the trials. The results are shown in Fig. 4.9 and given as the average of twenty trials. It can be seen from Fig. 4.9a that VIGP achieved the shortest average height of trees in the final population. However, in Fig. 4.9b, VIGP is seen to achieve poorer results in terms of the hit rate. One possible explanation for this result is that the functions only take two values (zero or one) in the 6-parity problem and never diverge. In the real-valued case, functions in a complicated tree form tend to be of high degree and exhibit complex behavior. As a result, such functions tend to be poor in terms of the MSE. In addition, when some of these eventually achieve a low MSE and proliferate in a population, it becomes difficult to find better solutions from them, because such complicated functions are sensitive to small modifications. That is, functions that are generated from complicated functions through a few crossovers or mutations have behavior that is as complex as that of the parents, but often have a large MSE. Therefore, they rarely take the place of their parents and soon disappear from the population. In other words, the evolution will be trapped in a local optimum. On the other hand, in the Boolean case of the 6-parity problem, even complicated functions take values of only zero or one and are thus not so sensitive to modifications. Therefore, better solutions can be found from them through the crossover or mutation operations. In order to support this discussion, we will demonstrate below that the solutions found by GP are rather complicated and not in the simplest form, while those by VIGP were in the simplest form.

We observed in Fig. 4.9b that VIGP achieves a lower hit rate than GP. However, when we closely compared the difference between the solutions found by VIGP and those of GP, we noticed that the form of the solutions was totally different. In the 6-parity problem, $a_1 + a_2 + a_3 + a_4 + a_5 + a_6 + 1$ can be considered as one of the simplest solutions, where a_i are variables. Here, we consider solutions not to be in the simplest form when they consist of more than six operators. As can be

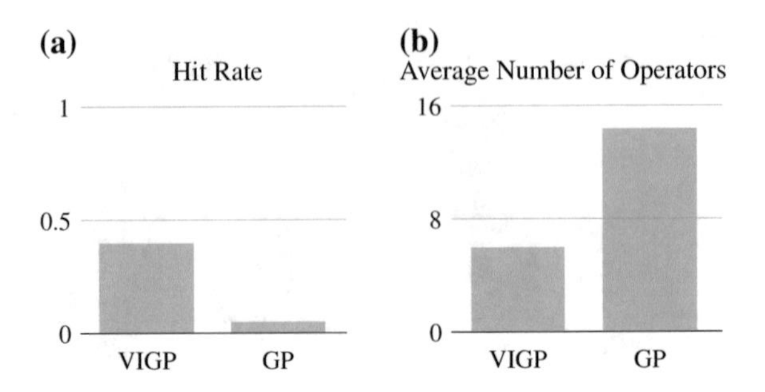

Fig. 4.10 Comparison between VIGP and GP with respect to the simplicity of solutions for the 6-parity problem. **a** Hit rate with respect to the simplest solution (i.e., $a_1 + a_2 + a_3 + a_4 + a_5 + a_6 + 1$). **b** Average number of operators in a solution. **a–b** The solutions found by VIGP were all in the simplest form, while those found by GP contained many more operators

seen in Fig. 4.10, the solutions found by VIGP are always the simplest, while those found by GP rarely are. When we compare the solutions in terms of the number of operators, we see that VIGP constructed solutions from about half the number of operators compared to GP (Fig. 4.10b). In summary, VIGP found the simplest solution considerably more frequently than GP.

4.3 The Kaizen Programming

In this section, we present the Kaizen programming (KP) introduced in [7–10]. It is a hybrid method for solving symbolic regression based on the Kaizen event [18] with the PLAN-DO-CHECK- ACT (PDCA) methodology [15]. In the real world, a Kaizen event is an event where experts propose their ideas and test them to tackle a business issue (see Fig. 4.11).

Many ideas are then combined to form a complete solution to the issue, which is known as the standard. During the Kaizen event, the PDCA methodology is employed to improve the quality of the current standard, where adjustments are planned, executed, checked, and acted. The cycle of PDCA is iterated until the business issue is solved. Since the contribution of each action at each cycle on solving the issue is analyzed and evaluated based on certain criteria, more knowledge on the issue is acquired. As a result, the experts learn from the improved information and will avoid useless or harmful actions and control the search process, thus resulting in a better solution. Figure 4.12 shows the flowchart of a high-level KP.

In order to solve regression problems, there are three basic modules which are important and necessary to KP according to [10]:

- Feature generation module, which is to propose ideas in the shape of regression features.
- Feature selection module, which is to evaluate and select contributive features.

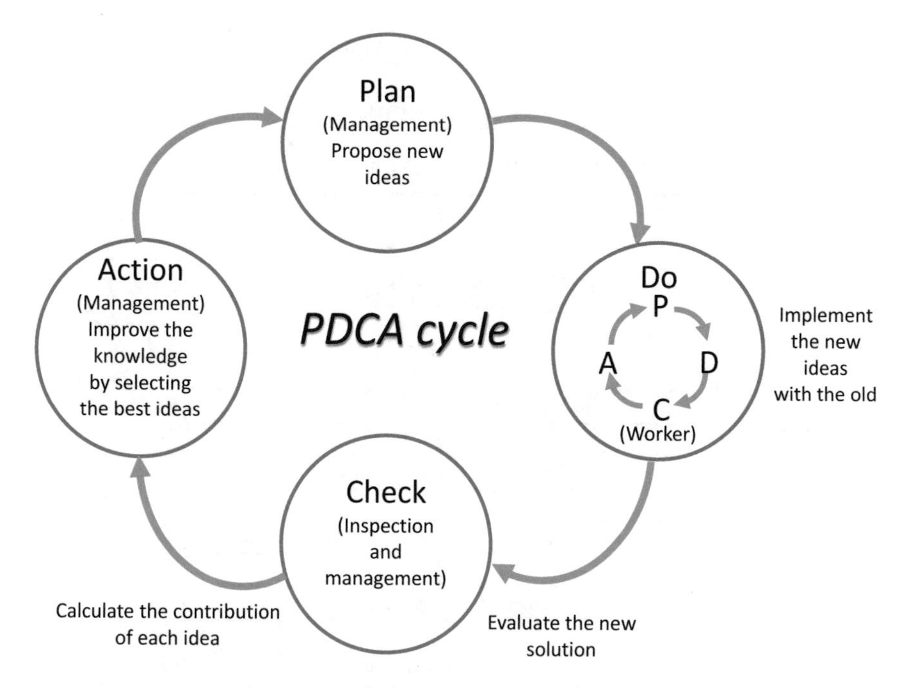

Fig. 4.11 Basic Kaizen programing flowchart

- Model generation module, which is to create a complete solution using generated features.

The feature generation module defines the actual form of the expert, including its data structure and executing procedures. Experts will propose new ideas to solve the problem, for example, by modifying some of the ideas contained in the current best solution. For solving regression problem, an idea can be a mathematical expression, for example, $\frac{1}{2}x^3 + \sin x$, which is equivalent to a feature used in a linear regression model.

At the feature selection module, new ideas are evaluated to generate an actual partial solution, which can be treated as an N-dimensional vector, where N is the number of training data. These partial solutions are combined with old ones existing in the current standard, resulting in a final complete solution composed of partial solutions. Each partial solution is then evaluated to indicate its contribution on solving the problem. Due to the relationship among those partial solutions, a measurement considers not only the independent quality of each partial solution, but also their dependency is used in this step. The major difference of conventional evolutionary algorithms with KP is that they work with individuals where each one is a complete solution. Therefore, in KP, individuals collaborate with each other, so the size of the population is smaller compared to conventional evolutionary algorithms, where individuals must compete with each other to survive.

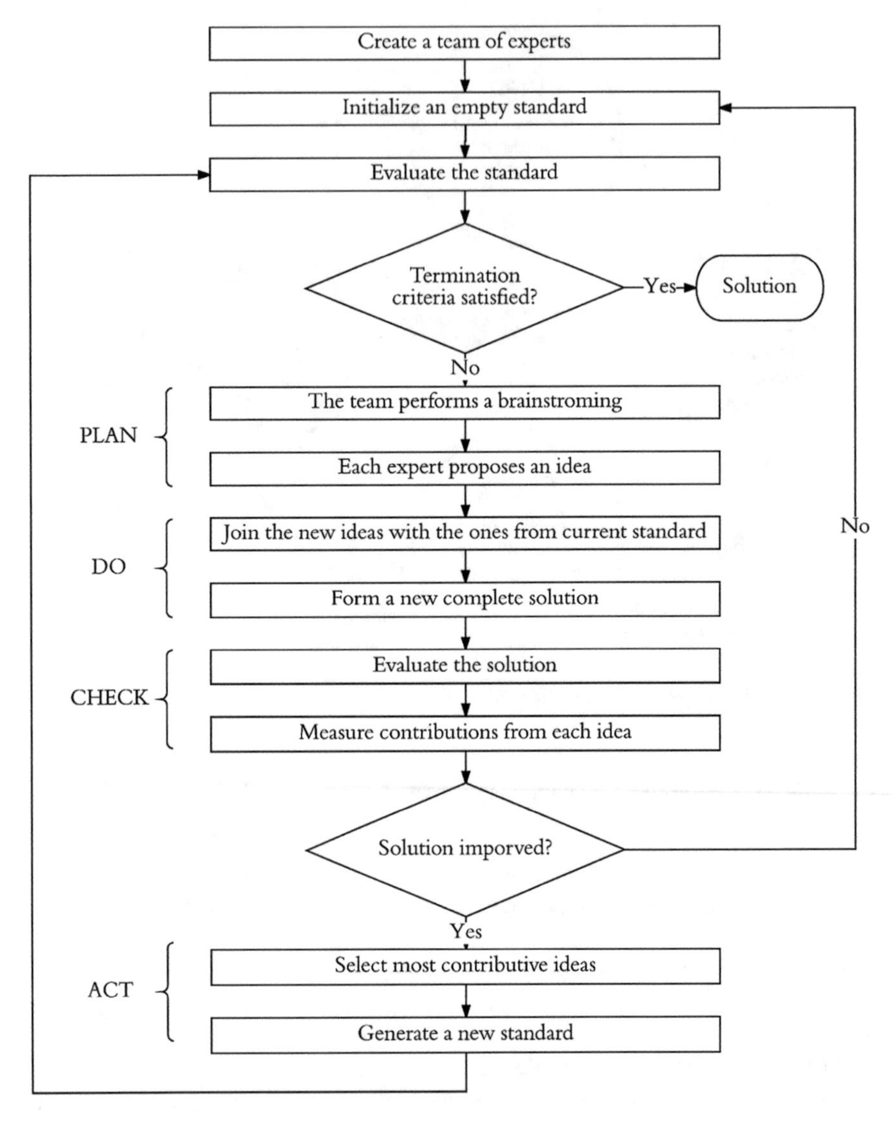

Fig. 4.12 Flowchart of a high-level KP

At the model generation model, the standard (complete solution) is generated based on the result in the last model. For example, after several most significant partial solutions are selected from the structure, the coefficients are fitted to the data by performing a maximum likelihood estimation (MLE, see Eq. (4.17) below), which is equivalent to minimize the root mean square error (RMSE). The evaluation of the quality of the complete solution is also needed. One of the reasons is to estimate if the algorithm gets stuck in a local optimum.

According to [10], the PDCA cycle plays the role for encouraging experts to propose ideas which are meaningful and useful for solving the problem. It is very important because the criteria for evaluating the contribution replace the place of fitness on the entire solution. Therefore, the employed methods to perform the second and third modules should be equivalent or at least logically connected to each other, an example is to use the same criteria to choose the features to construct the model, and more precisely, the method used to solve the problem and the method employed to evaluate the contribution of the partial solutions are equivalent. Provided that the feature selection module and the model generation model were unrelated, then the ideas meaningful to the procedure employed in the feature selection module would not to be meaningful to the method that finally solves the problem.

KP guides the search process to the goal based on the continuously increasing knowledge of the problem. It is thus necessary to use techniques to provide such knowledge. As a result, machine learning techniques or statistical approach must be employed efficiently for the hybrid structure of KP. Without the help from these techniques, KP reduces to a simple EC algorithm, where the search depends only on the selection pressure.

Here, we present a simple example of how KP is used for symbolic regression referenced from [10]. The KP implementation uses GP to provide features, uses MLE (maximum likelihood estimation, which is equivalent to minimize RMSE, see p. 129) to construct the test solution, and uses NHST (null hypothesis significance testing, see p. 128) to build the standard.

Assume the ground-truth target function is given by:

$$y = x^3 + x,\ x \in \mathrm{R}. \tag{4.14}$$

The regression objective is to predict output y from given input x. The assumption of the input–output pairs is given by:

$$\{x_n, t_n\}_{n=1}^{N} = \{(0.0, 0.0)\,,\ (0.1, 0.101)\,, \ldots,\ (1.0,\ 2.0)\}, \tag{4.15}$$

where the input is a uniform distributed sequence in the range between 0.0 to 1.0.

If we make an assumption that t is given by the summation of a constant term and a linear term, the linear regression model is given in the form:

$$y = w_0 + w_1 x, \tag{4.16}$$

where w_0 and w_1 are unknown parameters, and their values are to be estimated from the given data. This can be done by minimizing an error function. A general accepted error function used in the regression domain is the RMSE, given by:

$$RMSE[y] = \sqrt{\frac{1}{N}\sum_{n=1}^{N}(y(x_n) - t_n)^2}. \tag{4.17}$$

Table 4.4 Coefficients and p values for the example problem from [10]

ϕ_i	$coefficient(w_i)$	p value$_i$
1.0	~~1.18×10^{-16}~~	~~0.04~~
$\phi_1 = x^2$	~~2.89×10^{-13}~~	~~0.15~~
$\phi_2 = \sqrt{x}$	~~1.34×10^{-12}~~	~~0.14~~
$\phi_3 = \log(x+1)$	~~1.13×10^{-12}~~	~~0.15~~
$\phi_4 = x^3$	1.00	5.07×10^{-54}
$\phi_5 = x$	1.00	1.41×10^{-48}
$\phi_6 = \sqrt{\log(x+1)}$	~~-1.32×10^{-12}~~	~~0.14~~
p value > 0.05 means statistically $coefficient(w_i) \approx 0$.		
Ground-truth equation: $y(x) = x + x^3$		
I-O pairs: $\{(0.0, 0.0), (0.1, 0.101), \ldots, (1.0, 2.0)\}$		
Current population: ϕ_1, ϕ_2, ϕ_3		
Offspring produced: ϕ_4, ϕ_5, ϕ_6		
Test solution: $y_{test} = w_0 \times 1.0 + \sum_{i=1}^{6} w_i \phi_i$		
Updated solution: $y = 1.34 \times 10^{-16} + 1.0 \times x + 1.0 \times x^3$.		

Note that RMSE is a nonnegative quantity that would be zero if, and only if, the function y were to pass exactly through each training data point; thus, low error values are desired. By analytically finding the derivation of RMSE w.r.t the unknown parameters w_0 and w_1, and setting it to zero, we are able to determine the best approximation function to the ground-truth target:

$$y = -0.189 + 1.928x. \tag{4.18}$$

The fitness value of the resulting approximation function calculated using Eq. (4.17) is:

$$RMSE[y] = 0.1345. \tag{4.19}$$

We shall explain how to employ KP to solve this regression problem. For KP, a population of individuals is the foundation to evolve and to reach the final solution. Assume the current population of KP contains three individuals (see Table 4.4):

$$\{ind_i\}_{i=1}^{3} = \{x^2, \ \sqrt{x}, \ \log(x+1)\}. \tag{4.20}$$

The fitness values of these individuals calculated by RMSE are:

$$\begin{aligned} \{fitness\,[ind_i]\}_{i=1}^{3} &= \{RMSE\,[ind_i]\}_{i=1}^{3} \\ &= \{0.5188979, \ \ 0.4121862, \ \ 0.576697\}. \end{aligned} \tag{4.21}$$

For KP, we should first select several promising individuals and apply genetic operators on them to generate new offspring. Here, we assume the selection and production procedures are executed in advance, resulting in three offspring:

$$\{child_i\}_{i=1}^3 = \{x^3,\ x,\ \sqrt{\log(x+1)}\}. \tag{4.22}$$

Similarly, the fitness of them calculated in RMSE is:

$$\begin{aligned}\{fitness\,[child_i]\}_{i=1}^3 &= \{RMSE\,[child_i]\}_{i=1}^3 \\ &= \{0.591608,\ \ 0.4240932,\ \ 0.4862297\}.\end{aligned} \tag{4.23}$$

If we simply choose the best (with the smallest fitness value since it is the error) three individuals among the current population together with the newly generated offspring, the resulting next generation contains:

$$\begin{aligned}\{nxtInd_i\}_{i=1}^3 &= \{ind_2,\ child_2,\ child_3\} \\ &= \{\sqrt{\mathrm{x}},\ \ \mathrm{x},\ \ \sqrt{\log(\mathrm{x}+1)}\}.\end{aligned} \tag{4.24}$$

Unfortunately, this is not a desiring outcome since one of the correct partial solutions, i.e., $child_1 = x^3$, vanishes in the next generation. However, this expression obtained by $child_1$ might be generated again by GP system if enough time is given or if another selection mechanism is employed to keep it in the population. The issue of GP still exists, where an important part of the solution was eliminated in the evolutionary process via competition.

Fortunately, KP will preserve the desired $child_1$ since it works by the cooperation of partial solutions. Here, we assume that KP is using GP as the feature generation module.

Starting with the PLAN step (see Fig. 4.11), where KP should propose new ideas for constructing a new test solution, three individuals $\{ind_i\}_{i=1}^3$ are selected and modified by genetic operators to generate new features, resulting in $\{child_i\}_{i=1}^3$.

Next, the DO step will combine all individuals into a set of features, and it is common for regression problem to introduce a constant term (known as the bias) as a supplement, resulting in:

$$\begin{aligned}\{feature_i\}_{i=1}^7 &= \{ind_i\}_{i=1}^3 \cup \{child_i\}_{i=1}^3 \cup \{1\} \\ &= \left\{x^2,\ \ \sqrt{x},\ \ \log(x+1),\ \ x^3,\ \ x,\ \ \sqrt{\log(x+1)},\ \ 1\right\},\end{aligned} \tag{4.25}$$

The set of features are then evaluated on the given input data, and using the same technique of linear regression introduced in the beginning of this section to construct the test solution:

$$testSol = \sum_{i=1}^{7} w_i * feature_i, \tag{4.26}$$

where the unknown coefficients of features are estimated by minimizing RMSE. Their values are presented in Table 4.4, referenced from [10].

In the CHECK step, the test solution constructed in the DO step will be analyzed to evaluate the contribution or effectiveness of each feature in helping solve the regression problem. In KP, the evaluation of contribution is performed by the null hypothesis significance testing (NHST), which can be viewed as a procedure for feature selection.

The NHST used in KP is based on the student's t-distribution, which is obtained by mixing infinite numbers of Gaussian distributions with the same expectations but different variances. Compared to the Gaussian distribution, the student's t-distribution has an important property called robustness when used to describe the real-world data. The robustness means it is less sensitive to the presence of a few outlier data. The probability density function (pdf) of the student's t-distribution is given by:

$$\mathrm{St}\left(x \mid \mu,\lambda,v\right)=\frac{\Gamma\left(\frac{v}{2}+\frac{1}{2}\right)}{\Gamma\left(\frac{v}{2}\right)}\left(\frac{\lambda}{\pi v}\right)^{\frac{1}{2}}\left[1+\frac{\lambda\left(x-\mu\right)^2}{v}\right]^{-\frac{v}{2}-\frac{1}{2}}, \tag{4.27}$$

where $\Gamma(x)$ is the Gamma function and is defined by:

$$\Gamma(x)=\int_0^\infty y^{x-1}e^{-y}dy. \tag{4.28}$$

Since it is based on the students' t-distribution, the NHST used in KP is also known as the t-test. In order to evaluate the contribution of a feature in a linear regression model, t-test actually checks the significance of the corresponding coefficient of that feature. Specifically, for each coefficient w_i, two hypothesis statements of its significance are made:

$$\mathrm{H}_0{:}w_i=0, \tag{4.29}$$

$$\mathrm{H}_1{:}w_i\neq 0. \tag{4.30}$$

Next, the test will verify the reliability of each hypothesis statement. If H_0 is much more reliable, it means w_i is not significant and should be statistically treated as zero; thus, the corresponding $feature_i$ is multiplied to a zero coefficient, making it not contributive to the solution, and should be removed in the final standard. Obviously, vice versa.

Here, we should reformulate our discussion by introducing matrix notations for the sake of simplicity. If we take

$$\mathbf{t}=\left(t_1,\ \ldots,\ t_N\right)^{\mathrm{T}}, \tag{4.31}$$

$$\mathbf{w}=\left(w_1,\ \ldots,\ w_7\right)^{\mathrm{T}}, \tag{4.32}$$

$$\phi\left(x\right)=\left(feature_1,\ \ldots,\ feature_7\right)^{\mathrm{T}}, \tag{4.33}$$

$$\mathbf{\Phi}=\left(\phi^{\mathrm{T}}\left(x_1\right),\ \ldots,\ \phi^{\mathrm{T}}\left(x_N\right)\right)^{\mathrm{T}}, \tag{4.34}$$

then the MLE of unknown parameters is given by the following equation:

$$\mathbf{w}_{\text{ML}} = \left(\mathbf{\Phi}^{\text{T}}\mathbf{\Phi}\right)^{-1}\mathbf{\Phi}\boldsymbol{t}, \tag{4.35}$$

$$\beta_{\text{ML}}^{-1} = \frac{1}{N}\sum_{n=1}^{N}\{y\left(\mathbf{x}_n, \mathbf{w}_{\text{ML}}\right) - t_n\}^2, \tag{4.36}$$

where β_{ML} is the estimation of the precision of the Gaussian noise on the target variable. The MLE of the covariance matrix of the MLE result of coefficients is given by:

$$\begin{aligned}\mathbf{Cov}_{\text{ML}}\left(\mathbf{w}_{\text{ML}}\right) &= \mathbf{Cov}_{\text{ML}}\left(\left(\mathbf{\Phi}^{\text{T}}\mathbf{\Phi}\right)^{-1}\mathbf{\Phi}\boldsymbol{t}\right)\\ &= \beta_{ML}^{-1}\left(\mathbf{\Phi}^{\text{T}}\mathbf{\Phi}\right)^{-1}.\end{aligned} \tag{4.37}$$

For each coefficient w_i, we obtain the MLE of the corresponding variance of it:

$$\begin{aligned}var_{\text{ML}}\left(w_i\right) &= \left[\mathbf{Cov}_{\text{ML}}\left(\mathbf{w}_{\text{ML}}\right)\right]_{ii}\\ &= \left[\beta_{\text{ML}}^{-1}\left(\mathbf{\Phi}^{\text{T}}\mathbf{\Phi}\right)^{-1}\right]_{ii}.\end{aligned} \tag{4.38}$$

And the t statistics for the test is:

$$\left(t_0\right)_{w_i} = \frac{w_i}{\sqrt{var_{\text{ML}}\left(w_i\right)}}. \tag{4.39}$$

Thus, the p value of the coefficient w_i is derived by:

$$(p\text{-}value)_i = 2 \times \left(1 - T\left(|t_0|\right)\right), \tag{4.40}$$

where $T(x)$ is the cumulative distribution function (cdf) of the student's t-distribution.

The result of (p value)$_i$ verifies whether the coefficient w_i is statistically equivalent to zero. If $(p\text{-}value)_i > \alpha$ (α is the significant level of the hypothesis test, traditionally 0.05), then the hypothesis statement H_0 is more reliable, and thus, we should prune the corresponding $feature_i$ out of the model.

For the example problem, the estimated coefficients w_i and the computed (p value)$_i$ are listed in Table 4.4, referenced from [10].

The final step is ACT, where the contributive ideas should be selected and form a new standard. Obviously, in the table, only three features are contributive, which are the x, x^3, and 1. As a result, the reduced model after performing feature selection is given by:

$$reducedSol = 1.34 \times 10^{-16} + 1.0 \times x + 1.0 \times x^3 \tag{4.41}$$

Compared to the result found by GP, this solution has a better approximation to the ground-truth target function. If a better threshold (α, the significance level) of the hypothesis test is set (like 0.04), it might result in the exact form of the target. Though the target used in the example is relatively simple and may be found by GP after several generations, for KP it only requires a single iteration.

However, this example only considers the situation where the original standard (complete solution) is empty. Therefore, the new standard will simply be replaced by the result given in the form of Eq. (4.41). Provided that the original standard is not empty, there will be several differences to the presented example. First, the set of features will include other features from the original standard; second, the original standard will be replaced if, and only if, the $reducedSol$ has a higher quality. To evaluate the quality, a fitness function adjusted to the complexity of the model should be introduced, since $reducedSol$ may contain different numbers of features w.r.t the standard. In [7], the $\tilde{R}^2$ value (Eq. (4.52)) is used to compare the models without being mistaken by the number of features.

Another issue is that the PDCA methodology employed by KP results in a hill-climbing approach. Therefore, it is important to set criteria to restart the search for avoiding being stuck in a local optimum. The current standard should be preserved instead of being destroyed. The best standard discovered out of all restarts is returned as the global solution.

We believe that KP has several limitations requiring further improvement. The major limitations are listed in the following:

1. When the linear regression problem is solved by MLE, the singularity issue of the data variance matrix raises, which must be properly treated. The singularity is caused by similar or identical features used to construct the model, and it will result in the difficulty to estimate the parameters.
2. The hypothesis test requires a properly adjusted value for the significance level. If the threshold is too high, the complexity of the solution will increase excessively, and the redundancy in the solution will affect us to acquire inner knowledge of it. On the other hand, an overly low threshold will make it hard to find a good approximation to the target. Though, in the first case, we could control the excessive growth of the complexity by limiting the number of features used in the standard. It is still problematic to select a proper limit provided that we have no prior knowledge to the problem.

4.4 RVM-GP: RVM for Automatic Feature Selection in GP

This section introduces a new method to symbolic regression, based on the Sparse Bayesian kernel method, i.e., Relevance Vector Machine (RVM, see Sect. 2.2.3). We call this method RVM-GP [13, 14, 29], which integrates a GP-based search of tree structures, and a Bayesian parameter estimation employing automatic relevance determination (ARD). We extend the feature selection module of KP based on RVM. RVM-GP works without performing hypothesis test, thus requiring no prior knowledge to set the threshold. It also deals with singularity automatically.

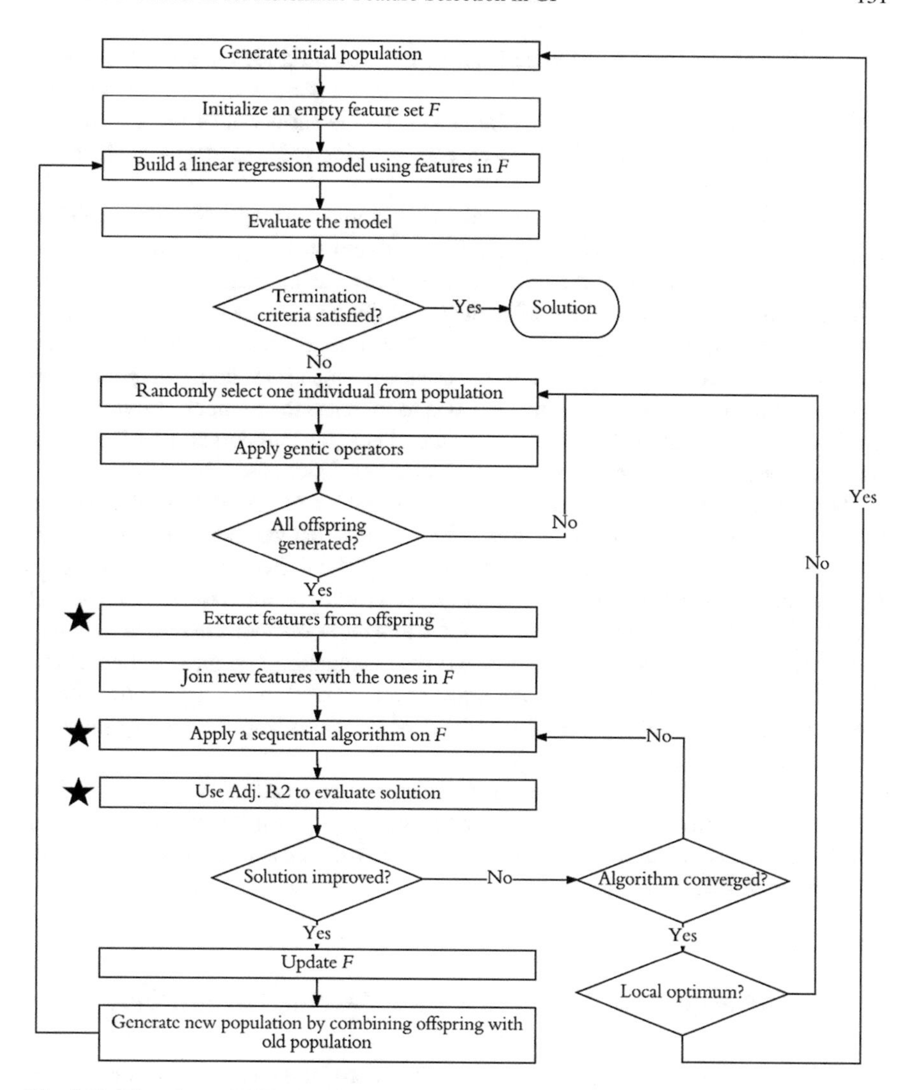

Fig. 4.13 Flowchart of RVM-based method

In conventional GP, the evolution process is done only by expecting that a new random solution presents a better fitness than the current solution, without the guide from deterministic approach. To overcome GP's major difficulty, we extend the work of KP in [7] by improving the feature selection with ARD, using RVM. Because RVM requires no prior knowledge to set a threshold and deals with the singularity issue automatically, RVM-GP also overcomes the limitations of KP. Figure 4.13 shows the basic flowchart of the method. In the following paragraphs, we will describe three key steps which are shown with a star mark ($\star$).

A sufficient number of candidate features must be produced in the first place to ensure the quality of the regression model. In the experiment, we found that the performance of RVM-GP is strongly influenced by the variety of features it utilizes. The lack of candidate features will decrease the growing speed of fitness and increase the risk of being stuck in local optimum.

As we no longer view the individuals as solutions, the primary duty of GP is to generate feature functions. Empirically, we found that a good feature function usually exists in a subtree of the individual. However, as the mapping from a tree to a function is very complex, the final behavior of the tree can be very different (e.g., $e^{\sin(x^2)}$ and x^2). If we simply use the entire tree as one feature function, then all the subtrees are wasted, and it is difficult for them to reappear in the future generation again. One approach for solving this issue is to increase the number of individuals, but it still brings more subtrees to be wasted. Thus, the proper solution is to do a traversal for each single tree to extract all its subtree functions.

The traversal ensures the maximum usage of GP trees. On the other hand, the variety of functions should also be noted. It is obvious that we should not use similar or identical functions as features. The training of RVM will be difficult to converge on repeated basis or even fail due to singularity. As a result, in practice a computational algebra library is used to transform a primitive tree into a symbolic expression, which will simplify the function (thus reduce the next evaluation complexity) and check the uniqueness of them. Furthermore, because an expression can appear frequently along the process, a LRU (Least Recently Used) cache is implemented to save the evaluation result of an evaluated expression.

In summary, it takes two steps for a GP primitive tree to construct candidate functions. First, the entire tree will be traversed to yield all its subtree functions. Second, all the subtrees will be transformed into symbolic expressions and filtered, resulting in different functions.

4.4.1 *The Sequential Sparse Bayesian Learning Algorithm*

After the construction of candidate feature functions, the next step is to put them into RVM for training. In practice, the sequential approach for solving the optimization problem is employed, which improves training speed significantly. Here, we briefly explain the mathematical background of it according to [33]. Continuing from Eq. (2.48), the problem is to determine the hyperparameters α and β [25, pp. 5–7]:

$$\boldsymbol{\alpha}^*, \beta^* = \underset{\boldsymbol{\alpha},\beta}{\operatorname{argmax}} \left(\ln p \left(\mathbf{t} \mid \mathbf{X}, \boldsymbol{\alpha}, \beta \right) \right), \tag{4.42}$$

where we make explicit all of the dependence of the marginal likelihood, i.e., Eq. (2.48), on each of the α_i and then optimize it explicitly. To do this, we first pull out the contribution from α_i in the matrix $\mathbf{C}$ defined by Eq. (2.49) to give the following equation:

$$\begin{aligned}\mathbf{C} &= \beta^{-1}\mathbf{I}+\sum_{j\neq i}\alpha_j^{-1}\varphi_j\varphi_j^{\mathrm{T}}+\alpha_i^{-1}\varphi_i\varphi_i^{\mathrm{T}} \\ &= \mathbf{C}_{-i}+\alpha_i^{-1}\varphi_i\varphi_i^{\mathrm{T}},\end{aligned} \tag{4.43}$$

where φ_i denotes the ith column of $\mathbf{\Phi}$. The matrix $\mathbf{C}_{-i}$ represents the matrix $\mathbf{C}$ with the contribution from basis function i removed. Using the Woodbury matrix identities,[2] the determinant and inverse of $\mathbf{C}$ can then be written as follows:

$$|\mathbf{C}| = |\mathbf{C}_{-i}|(1+\alpha_i^{-1}\varphi_i^{\mathrm{T}}\mathbf{C}_{-i}^{-1}\varphi_i), \tag{4.44}$$

$$\mathbf{C}^{-1} = \mathbf{C}_{-i}^{-1} - \frac{\mathbf{C}_{-i}^{-1}\varphi_i\varphi_i^{\mathrm{T}}\mathbf{C}_{-i}^{-1}}{\alpha_i+\varphi_i^{\mathrm{T}}\mathbf{C}_{-i}^{-1}\varphi_i}. \tag{4.45}$$

Using these results, we can then write the log marginal likelihood function Eq. (2.48) in the following form:

$$L(\boldsymbol{\alpha}) = L(\boldsymbol{\alpha}_{-i})+\lambda(\alpha_i), \tag{4.46}$$

where $L(\boldsymbol{\alpha}_{-i})$ is simply the log marginal likelihood with basis function φ_i omitted, and the quantity $\lambda(\alpha_i)$ is defined below:

$$\lambda(\alpha_i) = \frac{1}{2}\left(\ln(\alpha_i) - \ln(\alpha_i+s_i)+\frac{q_i^2}{\alpha_i+s_i}\right), \tag{4.47}$$

and contains all of the dependence on α_i. Here, we have introduced the two quantities:

$$s_i = \varphi_i^{\mathrm{T}}\mathbf{C}_{-i}^{-1}\varphi_i, \tag{4.48}$$

$$q_i = \varphi_i^{\mathrm{T}}\mathbf{C}_{-i}^{-1}\mathbf{t}. \tag{4.49}$$

where s_i is called the sparsity and q_i is known as the quality of φ_i. As we shall see, a large value of s_i relative to the value of q_i means that the basis function φ_i is more likely to be pruned from the model. The "sparsity" measures the extent to which basis function φ_i overlaps with the other basis vectors in the model, and the "quality" represents a measure of the alignment of the basis vector φ_i with the error between the training set values $\mathbf{t}$ and the vector $\mathbf{y}_{-i}$ of predictions that would result from the model with the vector φ_i excluded [2].

The stationary points of the marginal likelihood with respect to α_i occur when the following derivative is equal to zero:

---

[2]For all invertible matrices A, B, C, and D of correct sizes, $(A + BDC)^{-1} = A^{-1} - A^{-1}B(D^{-1} + CA^{-1}B)^{-1}CA^{-1}$ holds true.

$$\frac{d\,(\lambda\,(\alpha_i))}{d\,(\alpha_i)} = \frac{\alpha_i^{-1} s_i^2 - (q_i^2 - s_i)}{2\,(\alpha_i + s_i)^2} \tag{4.50}$$

There are two possible forms for the solution. Recalling that $\alpha_i \geq 0$, we see that if $q_i^2 < s_i$, then $\alpha_i \to \infty$ provides a solution. Conversely, if $q_i^2 \geq s_i$, we can solve for α_i to obtain the following equation:

$$\alpha_i = \frac{s_i^2}{q_i^2 - s_i}. \tag{4.51}$$

Note that this approach has yielded a closed-form solution for α_i for given values of the other hyperparameters. This leads to a practical algorithm for optimizing the hyperparameters that has significant speed advantages as shown in Algorithm 4.1.

4.4.2 Model Selection in RVM-GP

The principal disadvantage of RVM is that the training involves optimizing a non-convex function. During the training phase, many distinct models with different number of active functions will be generated. In general, a model with more active functions behaves better on training set. However, a simpler one is preferred if the difference of the training error between two models is small. And the complexity of RVM scales with respect to the number of active bases, in the experiments we found that the speed of RVM-GP is significantly influenced by the number of functions kept in the record. As a result, it is important to compare the models produced in the training phase and select the simplest model among those which exceed a measurement of quality.

In [7], the adjusted R^2 is used for model comparison, as it considers not only the fitness value but also the number of active functions compared to the number of training data. The adjusted R^2 (noted as $\tilde{R^2}$ hereafter) is calculated using the following formula:

$$\tilde{R^2} = R^2 - (1 - R^2) \times \frac{p}{N - p - 1}, \tag{4.52}$$

where N is the number of training patterns, p is the number of active functions in a model, and R^2 is the constant of determination which is given by:

$$R^2 = 1 - \frac{\sum_{n=1}^{N} (t_n - y_n)^2}{\sum_{n=1}^{N} (t_n - \bar{t})}, \tag{4.53}$$

where $\bar{t} = \frac{1}{N} \sum_{n=1}^{N} t_n$, and y_n is the nth entry of the output vector of a model.

Algorithm 4.1 Sequential Sparse Bayesian Learning Algorithm

Set β. Set all α_i to ∞.
Select an initial function φ_1.
Set s_1 and q_1 using Eqs. (4.48) and (4.49). ▷ Note that $\mathbf{C}_{-1} = \beta\mathbf{I}$.
Set α_1 using Eq. (4.51).
Evaluate the posterior of $\mathbf{w}$ using Eqs. (2.23)–(2.25.
while not convergence **do**
  Evaluate q_i and s_i for all functions using Eqs. (4.48) and (4.49).
  Select a candidate function φ_j.
  if $q_j^2 > s_j$ and $\alpha_j < \infty$ **then**
    Update α_j using Eq. (4.51).
  end if
  if $q_j^2 > s_j$ and $\alpha_j = \infty$ **then**
    Include φ_j to the model.
    Initialize its α_j using Eq. (4.51).
  end if
  if $q_j^2 \leq s_j$ **then**
    Remove φ_j from the model.
    Set $\alpha_j = \infty$.
  end if
  Update β using Eq. (2.30).
end while

In RVM-GP, the $\tilde{R}^2$ is used as the criterion to select the best model from the converged models. Compared to RMSE, the advantage of $\tilde{R}^2$ is that it is scale-free, w.r.t. the absolute target values of the problem, and the result of this statistic is in the range (0, 1) where $\tilde{R}^2 = 1.0$ means a perfect fit, $\tilde{R}^2 > 0.99$ means a high-quality model, and $\tilde{R}^2 > 0.95$ means a medium-quality model.

4.4.3 RVM-GP Performance

This section presents a comparison of RVM-GP against the conventional GP and the KP presented in the literature [7] to solve symbolic regression benchmark functions. The experiments demonstrate that RVM-GP can outperform them, providing high-quality solutions for both training and testing sets. Several *keijzer* benchmarks [24] for the first experiment and the *nguyen* benchmarks [11] for the second are defined in Table 4.5, which were chosen to compare RVM-GP with the results presented in [7]. In the table, c randomly sampled points from the uniform distribution confined in the range $[a, b]$ are noted as $U[a, b, c]$. Points confined the range $[a, b]$ with successively equal intervals c are noted as $E[a, b, c]$.

For comparison, RVM-GP and the other methods in the literature [7] were configured as shown in Table 4.6. Configurations regarding KP, GP 50, and GP 500 are referenced from [7]. The maximum generations are set to result in a balance of the total number of individuals used among all methods. To obtain statistically

Table 4.5 Benchmark functions as in [7]

Function	Training data	Testing data
Experiment I (referenced from [24])		
$keizer1 = 0.3x \sin(2\pi x)$	$E[-1, 1, 0.1]$	$E[-1, 1, 0.001]$
$keizer2 = 0.3x \sin(2\pi x)$	$E[-2, 2, 0.1]$	$E[-2, 2, 0.001]$
$keizer3 = 0.3x \sin(2\pi x)$	$E[-3, 3, 0.1]$	$E[-3, 3, 0.001]$
$keizer6 = \sum_{i=1}^{x} \frac{1}{i}$	$E[1, 50, 1]$	$E[1, 120, 1]$
$keizer7 = \ln x$	$E[1, 100, 1]$	$E[1, 100, 0.1]$
$keizer8 = \sqrt{x}$	$E[1, 100, 1]$	$E[1, 100, 0.1]$
$keizer9 = \text{arcsinh}(x)$	$E[1, 100, 1]$	$E[1, 100, 0.1]$
Experiment II (referenced from [11])		
$nguyen1 = x^3 + x^2 + x$	$U[-1, 1, 20]$	
$nguyen2 = x^4 + x^3 + x^2 + x$	$U[-1, 1, 20]$	
$nguyen3 = x^5 + x^4 + x^3 + x^2 + x$	$U[-1, 1, 20]$	
$nguyen4 = x^6 + x^5 + x^4 + x^3 + x^3 + x^2 + x$	$U[-1, 1, 20]$	
$nguyen5 = \sin(x^2)\cos(x) - 1$	$U[-1, 1, 20]$	
$nguyen6 = \sin(x) + \sin(x + x^2)$	$U[-1, 1, 20]$	
$nguyen7 = \log(x + 1) + \log(x^2 + 1)$	$U[0, 2, 20]$	
$nguyen8 = \sqrt{x}$	$U[0, 4, 40]$	
$nguyen9 = \sin(x) + \sin(y^2)$	$U[-1, 1, 100]$	
$nguyen10 = 2\sin(x)\cos(y)$	$U[-1, 1, 100]$	

convincing results, RVM-GP is executed 50 times in the first experiment and 100 times in the second experiment on each benchmark, which is the same as the other methods presented in the literature [7].

In the first experiment, the execution of each method will be terminated if the maximum generation is reached or the current global solution has a higher fitness than 0.99999. This experiment is based on *keijzer* benchmarks and is mainly for comparing the model quality (fitness adjusted to complexity of model) and computational complexity. The training results are shown in Table 4.7, and testing results are shown in Table 4.8. Results regarding KP, GP 50, and GP 500 are referenced from [7]. The number of function evaluations (NFEs) is shown for comparison of computational complexity.

For seven *keijzer* functions, the training results show that RVM-GP achieved the highest values of fitness for two functions (*keijzer* 2 and 3), with five ties to KP (*keijzer*1, 6, 7, 8, and 9) and 2 ties to GP (*keijzer* 6 and 8). However, the RMSE value of RVM-GP is larger in case of ties. RVM-GP found high-quality models (median of fitness >0.99 for six functions, whereas KP found five and GP (50 and 500) found models with such quality for only four functions.

Table 4.6 Parameter settings

Parameter	Value		
	RVM-GP & KP	GP 50	GP 500
Experiment I.			
Population size	8	50	500
Max. generations	2000	500	50
Tournament selection	1	3	
Crossover probability	1.0	0.9	
Crossover operator	One point		
Mutation probability	1.0	0.05	
Mutation operator	90% Uniform and 10% ERC	Uniform	
Max. depth	2	15	
Primitives	$+, \times, -\text{n}, \text{n}^{-1}, \sqrt{(\lvert\text{n}\rvert)}$		
Terminals	x, $\text{N}(\mu = 0, \sigma = 5)$		
Stopping criteria	Max. gen. or $\tilde{R}^2$>0.99999		
Trails	50		
Experiment II.			
Maximum number of node evaluations	100,000	Null	
Max. depth	8		
Primitives	$+, -, \times, \div$, sin, cos, exp, log		
Terminals	x, y, const		
Stopping criteria	Maximum number of node evaluations or Abs. error < 0.01 for all cases		
Trails	100		

RVM-GP performs as good as KP and GP for simple problems,[3] and it performs obviously better when dealing with complex functions (*keijzer* 2 and 3), where KP can find models with low or poor quality and GP can only find very poor results. For these functions, RVM-GP requires a much lower NFEs (median) than the other methods. It should be noted that, when dealing with very easy targets (*keijzer* 6, 7, 8, and 9), the models found by RVM-GP are very sparse and rapidly meet the stopping criteria (fitness > 0.99999), which explains the RMSE value of RVM-GP is larger as the $\tilde{R}^2$ rewards sparse models.

With respect to the testing results, we see that the minimum and median of RMSE of RVM-GP are somewhat larger than KP (*keijzer*1, 6, 7, 8, and 9). However, the maximum errors of RVM-GP are far smaller than KP in *keijzer*1, 2, 3, 7, 8, and 9, which explains that the results of RVM-GP are more stable than the others, in other

[3] *keijzer* 6, 7, 8, and 9 are easy targets. *keijzer*1 is relatively easier compared with *keijzer* 2 and 3.

Table 4.7 Training results using *keijzer* benchmark functions

Func.	Stat.	RVM-GP			KP			GP 50			GP 500		
		Fitness	RMSE	NFEs	Fitness	RMSE	NFEs	Fitness	RMSE	NFEs.	Fitness	RMSE	NFEs.
keijzer1	Max	1.00E+00	2.06E-05	11,214	1.00E+00	2.28E-02	32,260	8.30E-01	1.09E-01	25050	5.34E-01	8.99E-02	32,170
	Med	**1.00E+00**	**4.14E-05**	**5311**	1.00E+00	2.21E-04	9492	4.14E-01	8.97E-02	25050	4.14E-01	8.97E-02	30,450
	Min	1.00E+00	8.18E-05	532	1.00E+00	1.93E-06	328	1.40E-01	4.83E-02	25050	4.12E-01	8.00E-02	26,700
keijzer2	Max	1.00E+00	1.09E-02	16,588	1.00E+00	2.19E-01	32,380	8.17E-01	2.37E-01	25050	1.84E-01	2.34E-01	32,310
	Med	**9.98E-01**	**9.04E-03**	**6654**	8.99E-01	6.75E-02	32190	1.96E-01	2.16E-01	25050	1.39E-01	2.23E-01	30,550
	Min	9.95E-01	8.49E-04	1052	1.25E-01	5.42E-04	2880	3.44E-02	1.03E-01	25050	5.50E-02	2.17E-01	28,460
keijzer3	Max	9.92E-01	2.42E-01	17,870	9.13E-01	3.36E-01	32,260	5.39E-01	3.64E-01	25050	1.53E-01	3.58E-01	31,950
	Med	**9.81E-01**	**9.75E-02**	**14,811**	3.83E-01	2.65E-01	32,170	1.06E-01	3.44E-01	25050	7.05E-02	3.51E-01	30,300
	Min	9.56E-01	3.89E-02	11,130	1.26E-01	9.98E-02	32,100	0.00E+00	2.47E-01	25050	3.26E-02	3.35E-01	28,300
keijzer6	Max	1.00E+00	1.27E-04	104	1.00E+00	1.18E-14	14,510	1.00E+00	9.90E-01	51	1.00E+00	9.90E-01	501
	Med	**1.00E+00**	4.26E-04	**19**	**1.00E+00**	**9.22E-16**	56	**1.00E+00**	9.90E-01	51	**1.00E+00**	9.90E-01	501
	Min	1.00E+00	4.26E-04	7	1.00E+00	6.99E-16	56	1.00E+00	9.90E-01	51	1.00E+00	9.90E-01	501
keijzer7	Max	1.00E+00	6.83E-03	43	1.00E+00	1.78E-02	224	1.00E+00	9.26E-01	25050	9.99E-01	2.51E-01	32,480
	Med	**1.00E+00**	2.26E-03	**8**	**1.00E+00**	**4.61E-04**	80	9.98E-01	1.75E-01	25050	9.98E-01	1.81E-01	30,660
	Min	1.00E+00	1.32E-03	7	1.00E+00	2.96E-06	56	9.39E-01	6.07E-02	25050	9.96E-01	1.05E-01	28,590
keijzer8	Max	1.00E+00	6.08E-03	38	1.00E+00	1.34E-02	160	1.00E+00	0.00E+00	101	1.00E+00	0.00E+00	501
	Med	**1.00E+00**	5.61E-05	**7**	**1.00E+00**	3.50E-15	56	**1.00E+00**	**0.00E+00**	51	**1.00E+00**	**0.00E+00**	501
	Min	1.00E+00	1.03E-06	7	1.00E+00	7.91E-16	56	1.00E+00	0.00E+00	51	1.00E+00	0.00E+00	501
keijzer9	Max	1.00E+00	9.19E-03	22	1.00E+00	2.91E-02	248	1.00E+00	1.01E+00	25050	9.99E-01	7.70E-01	32050
	Med	**1.00E+00**	5.41E-03	**7**	**1.00E+00**	**1.59E-03**	88	9.99E-01	1.71E-01	25050	9.99E-01	1.35E-01	30,300
	Min	1.00E+00	8.08E-04	7	1.00E+00	9.77E-06	56	9.48E-01	8.47E-02	25050	9.69E-01	1.12E-01	25500

Table 4.8 Testing results using *keijzer* benchmark functions

Func.	Stat.	RVM-GP	KP	GP 50	GP 500
		RMSE			
keijzer1	Min	7.096E-03	1.067E-05	4.528E-02	7.876E-02
	Median	2.564E-02	**4.360E-04**	8.260E-02	8.199E-02
	Max	1.000E-02	5.737E+06	2.423E+03	1.304E+00
keijzer2	Min	9.769E-03	6.858E-04	1.729E-01	2.089E-01
	Median	6.755E-02	**6.849E-02**	2.391E-01	2.217E-01
	Max	1.566E+00	8.229E+00	2.345E+01	1.002E+00
keijzer3	Min	1.101E-02	9.673E-02	2.950E-01	3.412E-01
	Median	**5.022E-01**	2.065E+00	3.889E-01	3.526E-01
	Max	1.302E+01	9.984E+01	9.999E+08	2.710E+00
keijzer6	Min	6.867E-03	5.490E-16	9.958E-01	9.958E-01
	Median	9.712E-03	**8.808E-16**	9.958E-01	9.958E-01
	Max	2.215E-02	8.030E-14	9.958E-01	9.958E-01
keijzer7	Min	1.594E-03	3.888E-04	7.352E-02	1.063E-01
	Median	**1.828E-03**	1.272E-02	2.040E-01	1.711E-01
	Max	4.416E-02	4.491E+02	9.995E+09	5.502E+08
keijzer8	Min	1.029E-06	1.232E-15	0.0	0.0
	Median	1.777E-04	**6.077E-15**	**0.0**	**0.0**
	Max	2.679E-03	1.628E+00	0.0	0.0
keijzer9	Min	8.180E-04	4.739E-03	1.452E-01	1.393E-01
	Median	**4.545E-03**	1.026E-01	3.650E-01	1.513E-01
	Max	E- 9.635E-03	3.823E+09	9.995E+08	Inf.

words, more robust on the testing set. In practice, the robustness on testing set is much more important than a small difference in error.

In the second experiment, the termination condition is set to the maximum number of node evaluations reached. The termination will be considered as a fail, if in any one of the training cases, the absolute error is larger than 0.01. This experiment is based on *nguyen* benchmark functions. It uses the same interval for training and testing, but the sets of points are distinct. The result of RVM-GP is shown in Table 4.9, and the comparison with other methods is referenced and shown in Table 4.10.

Table 4.10 shows the averaged results of RVM-GP. RVM-GP as well as KP is successful in fitting target curves because the RMSE for training and testing remains low values for every test. In terms of the efficiency, we notice that the number of objective function and node evaluations also remain low values for every test, which means that RVM-GP is faster in finding the correct result. Part of the reason is the implementation of cache, but more importantly is the sparse kernel method employed by RVM-GP results in fast convergence.

Table 4.9 Results using *nguyen* benchmark functions

Func.	Maximum error	Raw fitness	RMSE training	Func. Eval	Node. Eval	RMSE training	Succ. runs
nguyen1	0.00283	0.0182	0.00112	21.14	20.63	0.05112	100
nguyen2	0.00479	0.0236	0.00135	32.63	27.56	0.10625	100
nguyen3	0.00399	0.0271	0.00172	45.73	36.68	0.83062	100
nguyen4	0.00406	0.0275	0.00174	33.41	54.32	0.49868	100
nguyen5	0.0006	0.004	0.0003	14.39	26.62	0.3641	100
nguyen6	0.00264	0.0188	0.00118	18.17	29.14	0.099	100
nguyen7	0.0022	0.0147	0.00093	8.07	14.04	0.28685	100
nguyen8	0.00145	0.0095	0.0006	17.26	29.66	0.1638	100
nguyen9	0.00352	0.0906	0.00114	75.29	177.5	0.70064	100
nguyen10	0.00442	0.102	0.00133	129.5	311.7	0.20869	100

Table 4.10 Number of successful runs using *nguyen* benchmarks

Method	F1	F2	F3	F4	F5	F6	F7	F8	F9	F10
RVM-GP	**100**	**100**	**100**	**100**	**100**	**100**	**100**	**100**	**100**	**100**
KP	**100**	**100**	**100**	**100**	**100**	**100**	**100**	**100**	**100**	**100**
ABCP	89	50	22	12	57	87	58	37	33	21
SC	48	22	7	4	20	35	35	16	7	18
NSM	48	16	4	4	19	36	40	28	4	17
SAC2	53	25	7	4	17	32	25	13	4	4
SAC3	56	19	6	2	21	23	25	12	3	8
SAC4	53	17	11	1	20	23	29	14	3	8
SAC5	53	17	11	1	19	27	30	12	3	8
CAC1	34	19	7	7	12	22	25	9	1	15
CAC2	34	20	7	7	13	23	25	9	2	16
CAC4	35	22	7	8	12	22	26	10	3	16
SBS31	43	15	9	6	31	28	31	17	13	33
SBS32	42	26	7	8	36	27	44	30	17	27
SBS34	51	21	10	9	34	33	46	25	26	33
SBS41	41	22	9	5	31	34	38	25	19	33
SBS42	50	22	17	10	41	32	51	24	24	33
SBS44	40	25	16	9	35	43	42	28	33	34
SSC8	66	28	22	10	48	56	59	21	25	47
SSC12	67	33	14	12	47	47	66	38	37	51
SSC16	55	39	20	11	46	44	67	29	30	59
SSC20	58	27	10	9	52	48	63	26	39	51

Table 4.10 compares results of RVM-GP with KP and other techniques working on the same curve-fitting problem.[4] We can see that RVM-GP and KP which build linear models outperform the other methods. And the performance of RVM-GP is as good as KP, both resulting in 100% successful runs.

In summary, RVM-based method avoids singularity and thresholding-required hypothesis test. It should be noted that RVM was conventionally used as a kernel technique though there is no restriction for applying it to any model expressed as a linear combination of feature functions.

4.5 PSOAP: Particle Swarm Optimization Based on Affinity Propagation

In this section, we show how to extend PSO in order to strike a good balance between local and global searches by using a clustering function. This is based on affinity propagation (AP) (see Sect. 2.2.8) to cluster particle swarms and update the positions of different particles on the basis of these cluster evaluations. We call this method as PSOAP (i.e., PSO+AP [32]).

The basic procedure is as follows. First, create clusters using AP clustering based on the position information for a particle swarm, allocating numbers to the clusters at random. Form a ring topology (Fig. 4.14) with these numbers arranged in order and use this as the topology for the clusters.

Next, find the cluster evaluation value Cb_k for each cluster (where k is the cluster number). The cluster evaluation value is defined as the optimum of the evaluation values for the particles belonging to the cluster. We introduce the neighborhood optimum $\vec{p}_n$, which is the optimum in the neighborhood of particle i. Here, the neighborhood refers to the particles in the same cluster as particle i or in the clusters linked to this cluster in the ring topology. For example, if the cluster number for the cluster containing particle i is j, then the best evaluation value out of Cb_{j-1}, Cb_j, and Cb_{j+1} is selected and its position producing this evaluation value is used for $\vec{p}_n$.

We modify Eq. (2.1) to (4.54) below for high-speed updates:

$$\vec{v}_i = \chi \cdot \left(\omega \cdot \vec{v}_i + \phi_1 \cdot (\vec{p}_i - \vec{x}_i) + \phi_2 \cdot \left(\frac{\vec{p}_g + \vec{p}_n}{2} - \vec{x}_i \right) \right) \tag{4.54}$$

Compared with traditional particle swarm optimization (PSO) using only the global best ($\vec{p}_g$), in PSOAP $\vec{p}_n$ also contributes to the high-speed update, and the algorithm can be expected to be correspondingly less likely to become trapped in a local optimum.

---

[4] Other techniques include Artificial Bee Colony Programming (ABCP), GP with standard crossover (SC), GP with no same mate (NSM), GP with context aware crossover (CAC), GP with soft brood selection (SBS), GP with semantic similarity-based crossover (SSC). See [20] for details.

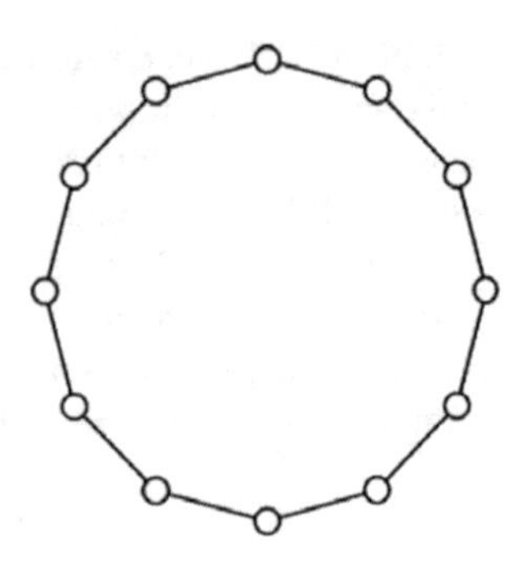

Fig. 4.14 Ring topology

In order to take advantage of the benefits of clusters, as well as introducing $\vec{p}_n$, a strategy for changing the parameter is used to set the rapid search on a cluster basis. Naturally, each cluster will have a different evaluation value, and so the relative importance of searching the cluster globally versus searching locally will vary according to these different evaluation values. For this reason, we vary the value of an inertia coefficient ω for each cluster according to this relative importance. If ω has a fundamentally large value, then the changes in velocity will be higher and so particles will be searched over a wider area. Conversely, a smaller value results in smaller changes in velocity and there will be a concentrated search for particles over a narrower range. Accordingly, based on these considerations, we set the minimum value for ω for each entire cluster as follows:

$$\omega_{base} = \omega_{max} - (\omega_{max} - \omega_{min}) \cdot \frac{g}{G}, \tag{4.55}$$

where ω_{min} and ω_{max} are the minimum and maximum for the base value of ω, while g represents the current search iteration and G is the maximum number of search iterations. In other words, ω_{base} decreases as the search progresses. On this basis, we use Eq. (4.56) to set the inertia coefficient ω for cluster j:

$$\omega_j = \omega_{base} + gap \cdot \frac{r_j - 1}{N - 1}, \tag{4.56}$$

where gap is the maximum increment for ω, while r_j is the rank for the evaluation value of cluster j among all clusters, and N is the total number of clusters. In other words, the higher the evaluation value for the cluster, the smaller ω becomes; conversely, the lower the evaluation value for the cluster, the larger ω becomes. By differentiating the values of ω in this way, we plan a good balance between local and global searches.

To evaluate the performance of PSOAP, we conducted numerical experiments using benchmark functions and compared the solution search performance of conventional PSO (with AP clustering) versus PSO with k-means clustering. We then compared the search performance against three non-PSO methods: the Firefly algorithm [37], differential evolution (see Sect. 2.1.2), and the Artificial Bee Colony algorithm [19]. For the benchmark functions, we used the following four functions listed in Table 4.18: Sphere function (F_1), Schwefel function (F_3), Rosenbrocks function (F_5), and Rastrigin function (F_9). The first two functions (F_1 and F_3) are unimodal functions, while the rest functions (F_5 and F_9) are multimodal.

The following parameters were fixed for all tests: number of updates (800), number of particles (25), and rank of the benchmark functions (10). The values of ϕ_1 and ϕ_2 were both fixed at 2.0, while the inertia coefficient ω decreased linearly from 0.7 to 0.4 as the number of updates increased. The number of clusters used with the k-means algorithm was fixed at 6. This value of 6 was the closest to the average value for the number of clusters generated as a result of clustering via AP for PSO as in PSOAP. To ensure the fairness of the comparison tests, the tests were conducted such that each function was evaluated 20,000 times.

The results of the experiment are shown in Table 4.11. Both PSO using the k-means algorithm and PSOAP had overwhelmingly better solution search performance than non-PSO methods. Moreover, although there was no clear difference between the performance with the k-means algorithm versus AP, overall, PSOAP showed better performance in terms of average evaluation values. This may be due to the fact that AP performs clustering using a number of clusters that matches the current state of the particles, and is more likely to perform more appropriate clustering than PSO using k-means. This makes clustering with AP less likely to become trapped in a poor local optimum. However, comparing this method with non-PSO methods, we see that although the performance of PSO is better against unimodal functions such as F_1 and F_2, the performance of PSO is worse than the other methods with respect to the multimodal functions. A possible reason why PSO does not have very good performance against multimodal functions may be because it is still becoming trapped in local optima despite the clustering. Clustering via AP does encourage efficient searches when particles are scattered, but once particles become bunched together, the clustering function no longer makes sense and the search stalls. Accordingly, it may be necessary to add a function that detects when particles have converged and responds.

The state of the cluster naturally differs by cluster, and according to this difference, the weighting of the necessity to search the cluster globally or locally changes. Therefore, changing the value of the parameter of the speed update formula for each cluster according to the state of the cluster is conceivable. Zhan et al. [38] proposed the parameter adjustment of the inertia coefficient ω and the acceleration constants ϕ_1 and ϕ_2. In this method, the parameters are adjusted by quantifying the search state by an evolutionary factor according to the dispersion state of the particles.

Table 4.11 Experimental results

Algorithm	Mean	Std	Min	Max
Sphere function (F_1)				
PSO	1.77E-16	3.17E-16	1.41E-18	1.20E-15
PSO with k-Means	1.55E-31	1.91E-31	8.11E-33	7.89E-31
PSOAP	1.28E-31	2.00E-31	3.63E-33	7.39E-31
Firefly	9.46E-06	2.43.E-05	4.25E-07	8.22E-05
DE	9.14E-19	2.55E-19	5.43E-19	1.42E-18
ABC	1.71E-16	5.35E-17	9.02E-17	2.53E-16
Schwefel function (F_3)				
PSO	4.32E-08	4.82E-08	7.52E-10	1.91E-07
PSO with k-Means	7.78E-12	1.16E-11	2.77E-15	4.14E-11
PSOAP	7.14E-12	1.08E-11	7.04E-14	1.65E-11
Firefly	8.89E-07	7.06.E-07	1.74E-07	2.62E-06
DE	6.15E-10	3.80E-10	7.41E-11	1.39E-09
ABC	5.75E+01	4.21.E+01	9.00.E+00	1.56.E+02
Rosenbrocks function (F_5)				
PSO	8.76E+01	1.20E+02	4.74E-01	3.33E+02
PSO with k-Means	6.70E+00	7,47E+00	2.70E-02	3.00E+01
PSOAP	5.76E+00	6.00E+00	1.51E-03	2.47E+01
Firefly	8.96E+00	1.44.E-02	8.92.E+00	8.98E+00
DE	3.99E-01	1.19E+00	1.65E-17	3.99E+00
ABC	1.45E-01	1.41.E-01	1.27.E-03	4.15.E-01
Rastrigin function (F_9)				
PSO	3.58E+00	2.27E+00	9.95E-01	1.10E+01
PSO with k-Means	2.29E+00	1.74E+00	1.66E-10	7.95E+00
PSOAP	1.84E+00	1.08E+00	0.00	2.98E+00
Firefly	8.66E-05	2.40.E-05	4.81E-05	1.41E-04
DE	9.95E-02	2.98.E-01	0.00	9.95.E-01
ABC	2.38E-09	7.01.E-09	1.78E-15	2.34E-08

4.6 Machine Learning for Differential Evolution

Differential evolution (DE, [30]) is a type of evolutionary algorithm that is used to solve optimization problems. DE is fast, robust, and easy to use; therefore, it is expected to be used to solve many optimization problems. However, the large number of fitness function evaluations encountered during solution searches is a disadvantage that becomes an issue in the practical use of DE. Speed is compromised when the cost per evaluation is large, and thus, the number of function evaluations should be reduced.

The origin of this problem lies in the design where function evaluation is conducted. The source of this problem is present in all generations during selection for all individuals that are generated by mutation and crossover. In other words, $N \times G$ function evaluations are necessary when there are N individuals in the population and G is the number of generations.

This is not a problem when the cost (time) of each fitness function evaluation is small. However, when $f(\vec{x})$ becomes complicated and the cost per evaluation increases, the effect that the large number of function evaluations has on the total DE cannot be neglected anymore. As a result, the DE solution search speed and performance decreases. The evaluation cost often becomes large in real applications because $f(\vec{x})$ is complicated and/or the dimension of $\vec{x}$ is large. In such cases, decreasing the number of function evaluations leads to the performance enhancement of DE because the effect of performance degradation from high evaluation cost is suppressed.

This section describes research on improving DE performance using machine learning. The problems to be optimized here are assumed to be sufficiently difficult, as mentioned above. This means that the time required to calculate the evaluation function is sufficiently longer than the time required in, for instance, classifier learning. Several of the applications under study are natural language processing and information search applications, in which large amounts of text data are processed [3], as well as robot motion planning data, for which simulations require a significant amount of time [16].

4.6.1 ILSDE

When generating the mutation vector in DE, $\vec{r}_1$, $\vec{r}_2$, and $\vec{r}_3$ are randomly chosen (see Eq. (2.5)). In other words, the mutation vector, which significantly affects the generation of the child vector, is derived from these randomly chosen individuals. Selecting these three individuals using a procedure that is superior to random selection will result in the generation of a superior mutation vector and a superior child vector. Therefore, the solution convergence speed should increase (locate a solution with in a fewer number of evaluations).

ILSDE (differential evolution based on improved learning strategy) [4] was developed based on the idea that DE performance can be improved by carefully choosing individuals in mutation. To be precise, a vector of three individuals, $\vec{r}_1$, $\vec{r}_2$ and $\vec{r}_3$, are selected based on the following strategy.

First, K individuals are randomly chosen to form a vector set X (K is an integer that is 3 or larger). The vector of $\vec{x} \in X$ that is most highly evaluated by the fitness function ($\vec{best}_1$) is chosen as $\vec{r}_1$, and the second-best vector ($\vec{best}_2$) is selected as $\vec{r}_2$. The vector with the worst evaluated value ($\vec{worst}$) is taken as $\vec{r}_3$.

Therefore, the mutation vector in ILDSE that corresponds to Eq. (2.5) in DE can be described as follows:

$$\vec{v}_i = \vec{best}_1 + F \cdot (\vec{best}_2 - \vec{worst}). \tag{4.57}$$

The difference vector of a bad ($\vec{worst}$) to good ($\vec{best}_2$) vector, which plays the role of the direction vector, is added to a vector with a high evaluated value at this point ($\vec{best}_1$). Thus, the generation of a mutation vector with a better evaluated value is expected.

The process after crossover is the same as standard DE, and thus, ILSDE is DE where Eq. (2.5) is replaced by Eq. (4.57). This change alone results in higher ILSDE performance compared to standard DE in many benchmark functions. Therefore, Eq. (4.57) generates better mutation vectors by using a better selection of individuals. This fact also demonstrates the importance that the selection of individuals ($\vec{r}_1$, $\vec{r}_1$ and $\vec{r}_3$) during mutation has on DE performance improvement.

Algorithm 4.2 ILSDE algorithm

1: Initialize(x)
2: **for** $g = 1$ to G_{max} **do** ▷ G_{max}:Maximum generations
3: **for** $i = 1$ to N **do** ▷ N:Population size
4: select K vectors at random ($\vec{x}_1, \vec{x}_2, \ldots, \vec{x}_K$)
5: $\vec{best}_1 = \mathrm{argmin}_{\vec{x}_i}\{f(\vec{x}_1), f(\vec{x}_2), \ldots, f(\vec{x}_K)\}$
6: $\vec{best}_2 = \mathrm{argmin}_{\vec{x}_i \neq \vec{best}_1}\{f(\vec{x}_1), f(\vec{x}_2), \ldots, f(\vec{x}_K)\}$
7: $\vec{worst} = \mathrm{argmax}_{\vec{x}_i}\{f(\vec{x}_1), f(\vec{x}_2), \ldots, f(\vec{x}_K)\}$
8: $\vec{v}[i] = \vec{best}_1 + F \times (\vec{best}_2 - \vec{worst})$ ▷ Mutation
9: DE-Crossover
10: DE-Selection
11: **end for**
12: **end for**

4.6.2 SVC-DE

SVC-DE [35] is a method that utilizes classification based on an SVM (see Sect. 2.2.2). Function evaluation in DE is carried out to compare parent and child vectors during selection, as is evident from Eq. (2.4).

Function evaluation is necessary even if the evaluated value of the child vector is worse than the parent vector, or if the parent vector remains unchanged in the next generation. Before function evaluation, if we can know whether the evaluated value of the child vector will be better or worse than the parent vector, the number of function evaluations can be reduced by preventing the parent vector from being subjected to function evaluation.

SVC-DE is based on this idea and conducts a prediction of the evaluated value using a classifier (SVM). The general flow of this algorithm is shown below. The process at the mutation and crossover phases is the same as standard DE (except for the maneuver used to obtain training examples).

Step1 Standard DE is carried out for a certain number of generations to accumulate vector $\vec{x}$ and the evaluated value of $f(\vec{x})$ as training data for learning.

Step2 During selection, find k training examples closest to a child vector, $\vec{u}_i$, among those accumulated in **Step1**.

Step3 Determine the positive and negative examples from the k training examples. The evaluated value, $f(\vec{x}_i)$, of parent vector $\vec{x}_i$ (of child vector $\vec{u}_i$) is compared against the evaluated values of the extracted k training examples. Examples where $f(\vec{x}_i)$ is a better or worse evaluated value are labeled as a positive or negative example, respectively.

Step4 For the k training examples labeled in **Step3**, $\vec{u}_i$ is defined as a positive example if all learned examples are positive examples, or it is labeled as a negative example if all training examples are negative examples. If there are both positive and negative examples among the k examples, the SVM is used to classify whether $\vec{u}_i$ is a positive example or a negative example.

Step5 If $\vec{u}_i$ is judged as a positive example, the value of function $f(\vec{u}_i)$ is evaluated, compared with the evaluated value of the parent vector, and the better vector is used in the next generation (the same procedure as standard DE). In addition, $(\vec{u}_i, f(\vec{u}_i))$ is accumulated as a new training example. If $\vec{u}_i$ was judged as a negative example, nothing is done and parent vector $\vec{x}_i$ remains unchanged in the next generation.

The basic principle behind this method is that machine learning (using a classifier) is introduced in the selection phase to predict and classify whether a child vector should be evaluated prior to function evaluation. Reducing the number of function evaluations through this method successfully improved the performance of DE compared to ILDSE.

Lu et al. [23] compared the performance of SVC-DE against other methods for 10 benchmark functions (see Sect. 4.6.3 for the detailed definition). The compared methods included standard DE, support vector regression DE (SVR-DE) (which utilizes regression-based selection), and ranking SVM-DE (RankSVM-DE) (which utilizes ranking-based selection). SVR-DE estimates actual evaluated values and compares individuals based on estimated values instead of simply determining whether an individual is superior to another. RankSVM-DE ranks the parent individual, the child individual, and nearby individuals and uses the ranking to judge whether the parent individual is superior to the child individual.

The experiment conducted by Lu et al. [23] assigned the SVC-DE parameters shown in Table 4.12. Table 4.13 compares SVC-DE with other methods. The used benchmark functions are described in Table 4.18. As can be seen from the table, SVC-DE is significantly superior in nine, six, and eight functions compared to DE, SVR-DE, and RankSVM-DE, respectively. Moreover, the performance of SVC-DE is approximately the same or better than other methods for almost all functions.

As an example, Fig. 4.15 shows the transition of the evaluated value from benchmark function F_4. The horizontal axis represents the number of times that the fitness function was called, and the vertical axis represents the logarithm of the evaluated value. The objective here is to find the minimum value of the function; a small value

Algorithm 4.3 SVC-DE algorithm

```
1: Initialize(x⃗)
2: for g = 1 to G_c do                              ▷ G_c: Generation size for collection
3:     do Mutation - Crossover - Selection
4:     Archive all (x⃗_i; f(x⃗_i)) into DB
5: end for
6: for g = G_c to G_max do                          ▷ G_max: Maximum generations
7:     do Mutation - Crossover
8:     for each u⃗_i do                                                 ▷ Selection
9:         NB =Neighborhood(u⃗_i, DB, k)                  ▷ DB: data (x⃗; f(x⃗)) set
10:        switch NB do                          ▷ NB: k training data (x⃗; f(x⃗))
11:            case all positive
12:                assert(class = +1)
13:            case all negative
14:                assert(class = −1)
15:            case mixed
16:                assert(class = SVM(u⃗_i, NB))                ▷ Neighborhood of u⃗_i
17:        if class == +1 then
18:            Archive (u⃗_i; f(u⃗_i)) into DB
19:            if f(u⃗_i) < f(x⃗_i) then
20:                x⃗_i = u⃗_i
21:            end if
22:        end if
23:    end for
24: end for
```

Table 4.12 SVC-DE parameters

D	30
NP	100
F	0.5
CR	0.9
G_{max}	500
G_c	20
k	40

means that the score is better. SVC-DE converges to a smaller value with a smaller number of evaluations after the few generations necessary to build a database are created.

The performance of SVC-DE is better than other methods because it is more suitable for the DE algorithm, in which the parent and child are compared pairwise. In other words, classification exhibits better learning efficiency than regression and ranking in DE selection.

One issue in SVC-DE is the small amount of training data available. There is no training data when searching is initiated, and therefore, training data must be accumulated while conducting the same process as standard DE. After a database is built, k closest data (which will be used as training data) are chosen from the database consisting of, at most, $NP \times G$ data. Newly generated child individuals in DE have

Table 4.13 SVC-DE versus DE, SVR-DE, and RankSVM-DE (w: win, d: draw, l: lose)

	versus DE	vsersus SVR-DE	versus RankSVM-DE
F_1	w	w	l
F_2	w	w	w
F_3	w	w	w
F_4	w	w	w
F_5	w	d	w
F_6	w	w	w
F_7	w	w	w
F_8	d	d	d
F_9	w	l	w
F_{10}	w	d	w

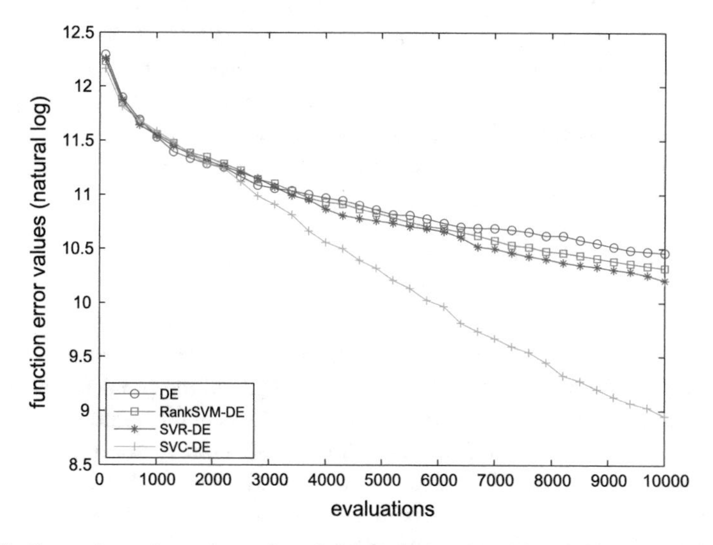

Fig. 4.15 Comparison of experimental result for F_4 (Reproduced from [23])

a smaller evaluated value compared to the parent individual in quite a few cases, and this tendency increases as the searches progress.

Positive examples where the child individual is superior to the parent cannot be sufficiently provided in SVC-DE. Thus, learning must be performed using a small amount of training data. Moreover, the amount of data in the database is small during the early stages of a search, and data are thought to be sparsely distributed in the search space. As a result, data that are not close enough may be selected as training data, and the quality of learning may be low.

Therefore, next section discusses a more accurate classification method for DE and attempts to achieve an efficient search mechanism.

4.6.3 TRADE: TRAnsfer Learning for DE

Sufficient training data for classifier learning are not guaranteed in SVC-DE, as proposed by Lu et al. [23]. However, increasing the number of samples used in learning may improve the performance of the classifier. This section describes a novel method called TRADE (TRansfer learning for DE). The goal of TRADE, as in SVC-DE, is to use a classifier that predicts the relative superiority between individuals during selection in DE to reduce the number of evaluation function calculations, and to optimize efficiency.

TRADE incorporates the idea of transfer learning during the learning process of the classifier, which predicts the relative superiority between individuals in DE. There are not many cases where a newly generated child individual in DE has a better evaluated value than the parent individual, and this tendency increases as a search progresses. Therefore, positive examples where the child individual is superior to the parent cannot be sufficiently provided in SVC-DE. Thus, learning has to be executed using a small amount of training data. However, prior research claims that the effectiveness of transfer learning improves when the number of samples in the target domain is fewer, which suggests a good fit for SVC-DE.

The feature vector of the classifier in SVC-DE is the parent vector ($\vec{x}$) with the candidate child vector ($\vec{u}$), i.e.,

$$V = (x_1, \ldots, x_D, u_1, \ldots, u_D), \tag{4.58}$$

where $\vec{x} = (x_1, \ldots, x_D)$ and $\vec{u} = (u_1, \ldots, u_D)$. However, it is almost impossible to use the DE vector value as it is. Transfer learning does not work well unless the landscapes of the problem to be optimized (f_t), and the problem provided for transfer learning (f_s), are very similar or the coordinates of the optimum solution are at approximately the same position. Data transferred from f_s not only degrade the performance of the classifier that predicts the relative superiority between individuals in f_t and conducts meaningless classifications, but also lead to negative transfer, which decreases the overall search efficiency.

Therefore, TRADE uses the first derivative of the fitness function as knowledge to be transferred from f_s, i.e.,

$$V = \left(\frac{\partial f}{\partial x_1}, \ldots, \frac{\partial f}{\partial x_D}, \frac{\partial f}{\partial u_1}, \ldots, \frac{\partial f}{\partial u_D} \right). \tag{4.59}$$

However, the derivative of the evaluation function is rarely known. Moreover, the cost to calculate the derivative is generally higher than the cost to calculate the evaluation function. As a result, the derivative to be used is estimated from nearby data. The differential at the position of the child individual vector is approximated from parent individual vectors near the target child individual vector. The simplest approximation method involves obtaining a D-dimensional hyperplane from D nearby vectors and using its coefficients as the value of the derivative.

Using only the differential as training data may decrease the learning efficiency in f_t compared to SVC-DE; thus, the vector placed next to the differential is also considered as the training data in learning, i.e.,

$$V = \left(x_1, \ldots, x_d, u_1, \ldots, u_D, \frac{\partial f}{\partial x_1}, \ldots, \frac{\partial f}{\partial x_D}, \frac{\partial f}{\partial u_1}, \ldots, \frac{\partial f}{\partial u_D} \right). \tag{4.60}$$

The TRADE algorithm is basically similar to SVC-DE, but the idea of transfer learning is incorporated to increase the number of samples that the classifier can use when learning. The overall flow of TRADE is summarized as follows:

1. Prepare a different problem (f_s) other than the problem to be optimized (f_t), conduct optimization using DE, and collect data for learning during DE.
2. Perform the same optimization of f_t as that performed in SVC-DE.
   (a) Database building: the first few generations are processed in a similar manner to standard DE in order to collect data for learning. Searching by a classifier is possible from the second generation by using training data from f_s and a small amount of training data from f_t.
   (b) Searching by a classifier: SVM classifier learning is conducted after a certain amount of data is accumulated from f_t and f_s, and searching is performed using a prediction of the relative superiority between individuals. A search will stagnate when the classifier judges all examples as negative examples, and thus, a certain ratio of examples predicted to be negative are evaluated.

The TRADE algorithm is shown in Algorithm 4.4. The addition of lines 1 and 14 represents the main difference between the TRADE and SVC-DE algorithms. Line 1 is the process that collects learning data transferred from f_s, and line 14 is the process that adds those from f_s to the data used in learning. Here, the data from f_t are those where the Euclidean distance between vector values are shorter, and the data from f_s have short Euclidean distances between differentials. Increasing the amount of data used in learning is expected to improve the performance of the classifier and reduce the number of unnecessary evaluation function calculations.

Again, we used the benchmark functions listed in Table 4.18. The TRADE parameters are the same as those applied in SVC-DE and are listed in Table 4.14. The amount of training data transferred from f_s, i.e., k_s, has to be adjusted using preliminary experiments. We used the RBF kernel in SVM learning (see Eq. (2.16)).

Table 4.14 TRADE parameters

D	30	$Emax$	10,000
NP	100	$MaxG_{DB}$	20
F	0.5	k	40
CR	0.9	k_s	40

Algorithm 4.4 TRADE algorithm

1: Collect source database DB_s =DE($MaxG_s$)
2: $g = 0$, $eval = NP \times MaxG_{DB}$, DB = emptyDB ▷ Database building with f_s.
3: Initialize a population $pop_g = \{\vec{x}_{i,g} \mid i = 1, \ldots, NP\}$
4: Archive all $(\vec{x}_{i,g}, f_t(\vec{x}_{i,g}))$ into DB
5: **while** $g < MaxG_{DB}$ **do**
6: $pop_{g+1} = \{\vec{x}_{i,g+1} \mid i = 1, \ldots, NP\} \Longleftarrow$Mutation_Crossover_Selection(pop_g)
7: Archive all $(\vec{x}_{i,g+1}, f_t(\vec{x}_{i,g+1}))$ into DB
8: $g = g + 1$
9: **end while**
10: **while** (evaluation count < $Emax$) **do** ▷ Search for f_t space.
11: $upop_{g+1} = \{\vec{u}_{i,g+1} \mid i = 1, \ldots, NP\} \Longleftarrow$ Mutation_Crossover(pop_g)
12: **for** each $\vec{u}_{i,g+1}$ **do**
13: NB_i =Neighborhood($\vec{u}_{i,g+1}, DB, k$)
14: NB_i+ =Neighborhood($\vec{u}_{i,g+1}, DB_s, k_s$)
15: **if** SVC($\vec{u}_{i,g+1}, \vec{x}_{i,g}, NB_i$)== +1 **then**
16: Archive $(\vec{u}_{i,g+1}, f_t(\vec{u}_{i,g+1}))$ into DB
17: **if** $f_t(\vec{u}_{i,g+1}) < f_t(\vec{x}_{i,g})$ **then**
18: $\vec{x}_{i,g+1} = \vec{u}_{i,g+1}$
19: **else**
20: $\vec{x}_{i,g+1} = \vec{x}_{i,g}$
21: **end if**
22: **else**
23: $\vec{x}_{i,g+1} = \vec{x}_{i,g}$
24: **end if**
25: **end for**
26: $pop_{g+1} = \{\vec{x}_{i,g+1} \mid i = 1, \ldots, NP\}$
27: $g = g + 1$
28: **end while**

As a numerical experiment to demonstrate the performance of TRADE, two functions, f_t and f_s, are chosen from benchmark functions and DE simulations were conducted for the following combinations:

1. Single-peak function f_t and single-peak function f_s
2. Single-peak function f_t and multipeak function f_s
3. Multipeak function f_t and multipeak function f_s

Standard DE and SVC-DE search in f_t only, and TRADE searches in f_s first by using DE to obtain training data for transfer learning before searching in f_t. The performance is compared by observing how the evaluated value transitions with the number of evaluation function calculations.

Figure 4.16 shows the results of a TRADE simulation where knowledge is transferred from one single-peak function to another. In the simplest example, F_1 and F_2 are chosen as f_s and f_t, respectively. Five simulations are conducted, and the average is shown. There is no significant difference between searches with TRADE, SVC-DE, and DE according to the graph, but the results of TRADE are slightly better from the middle to end of the search. Indeed, the minimum of the function after 10,000 evaluations is

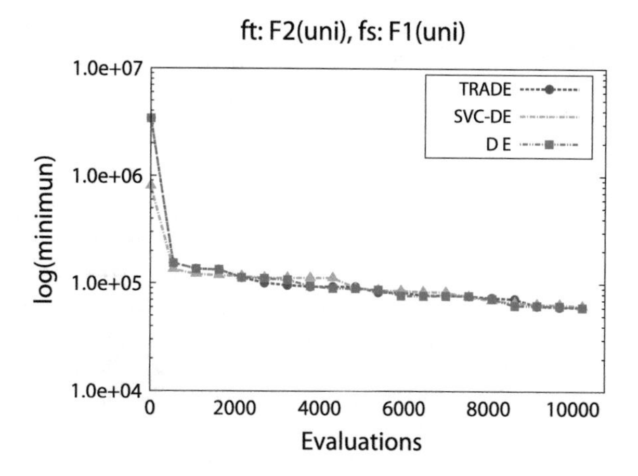

Fig. 4.16 Transition of evaluated value $(f_t, F_2; f_s, F_1)$

$$\text{TRADE}(5.286e+04) < \text{DE}(5.758e+04) < \text{SVC-DE}(6.195e+04)$$

thus, TRADE showed the best performance.

Poor classifier performance may be a reason why search performance does not improve so much. Therefore, classification accuracy is investigated. Positive examples are when the child individual is superior to the parent individual, and negative examples are when child individual is inferior to the parent individual. Binary classification results can be categorized into the following categories.

- **True Positive** (TP): Actually positive examples that are correctly predicted as a positive example,
- **True Negative** (TN): Actually negative examples that are correctly predicted as a negative example,
- **False Positive** (FP): Actually negative examples that are incorrectly predicted as a positive example,
- **False Negative** (FN): Actually positive examples that are incorrectly predicted as a negative example.

Figures 4.17 and 4.18 show the classifier performance in TRADE and SVC-DE, respectively. SVC-DE predicts eight to more than nine out of ten cases as negative examples. On the other hand, TRADE has a tendency to predict more positive examples in early stages when f_t is F2 or F6, but otherwise almost all cases are predicted as negative examples, as seen in SVC-DE.

TRADE basically calculates the evaluation function when the classifier predicts a positive example, as in SVC-DE. Therefore, a classifier that mostly predicts negative examples results in an extremely small number of evaluated individuals; this causes the search to stagnate.

As an intrinsic feature of DE, few child individuals are superior to the parent individual (positive example), and thus, the number of positive and negative examples in training data is unbalanced. TRADE collects nearby positive and negative examples

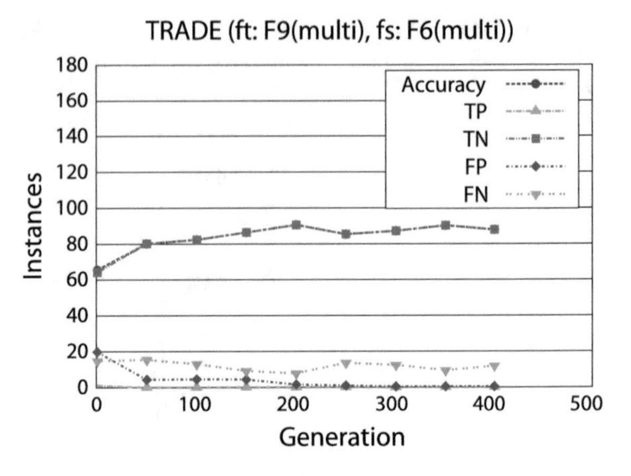

Fig. 4.17 Classification performance of TRADE in each generation (f_t, F_9; f_s, F_6)

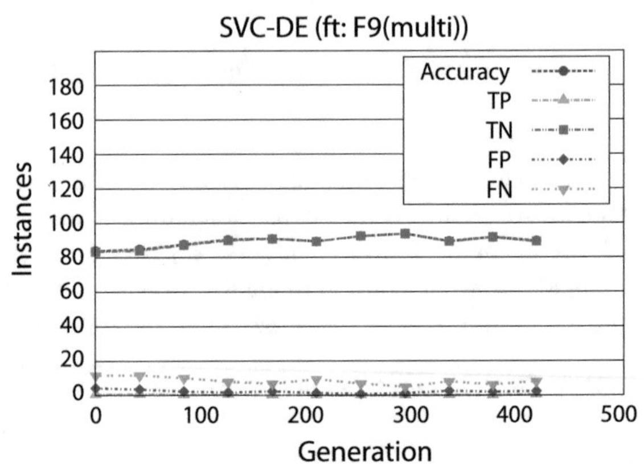

Fig. 4.18 Classification performance of SVC-DE in each generation (f_t, F_9)

to be classified based on the feature vector and Euclidean distance, and then conducts learning based on the same number of positive and negative examples. However, this method is not always the best, and the classifier learned to predict many negative examples.

One possible approach to solve this issue is to place the center of gravities of positive and negative examples at the same position. First, the center of gravity of nearby positive examples is obtained, and negative examples are collected such that the centers of gravity are similar. This will resolve the distance imbalance and is also expected to resolve the classification bias toward negative examples. Moreover, there are methods available to evaluate or bias the evaluation function [22].

The next section discusses the dynamic updating of training examples and the expansion of the classifier as alternative approaches.

4.6.4 NENDE: k-NN Classifier for DE

Improvement of the classifier precision is directly linked to the improvement of DE performance, as shown in the previous section. An increase in the accuracy rate of positive and negative example classification in individual selection results in a lower number of function evaluations necessary for convergence of the solution. Individuals that the classifier predicted to evolve (positive examples) are subject to crossover and subsequent processes, but not in those predicted to not evolve (negative examples). This means that search performance does not improve if classification is not appropriate.

Therefore, this section explains a method for enhancing classification accuracy. To be precise, improvement in classification methods using machine learning is employed in individual selection during mutation. Individual selection is carried out using a classifier based on the k-nearest neighbors (k-NN) algorithm, and hence, this method is denoted as NENDE (k-NN classifier for DE).

The rough flow of the overall algorithm is described as follows.

Step1 Standard DE is performed for a certain number of generations to accumulate training examples for the classifier.

Step2 After accumulation, the classifier is used to select individuals at mutation (i.e., $\vec{r}_1$, $\vec{r}_2$, and $\vec{r}_3$) and classifies them into positive and negative examples.

Step3 The set of selected individuals judged as positive examples proceed to crossover and subsequent processes. No further processing is conducted for individuals judged as negative examples.

Step4 Return to **Step2** after selection. This loop is carried out until the number of function evaluations reaches a predetermined number.

This algorithm employs a classifier to reduce the number of function evaluations by removing selected individuals ($\vec{r}_1$, $\vec{r}_2$, and $\vec{r}_3$) where the evaluated value of the generated child vector will be lower.

NENDE is similar to SVC-DE, but a significant difference is that a classifier is used in the mutation phase in this algorithm. Using a classifier at the individual selection phase instead of predicting the evaluated value of the generated child vector is expected to generate a better individual selection method, increase the solution search speed, and further improve performance.

The most important point of NENDE is the classification accuracy of the classifier. Bad classification regarding whether the selected individuals are good or not would not generate good child vectors and does not lead to increased solution search speed. The following three points must be considered to improve classification accuracy:

- Classification method,
- Method to accumulate training examples,
- How to take the feature vector.

Providing appropriate training examples is very important for machine learning. The same training examples can be continuously used once established if the classification criteria do not change during learning. This does not work well in DE.

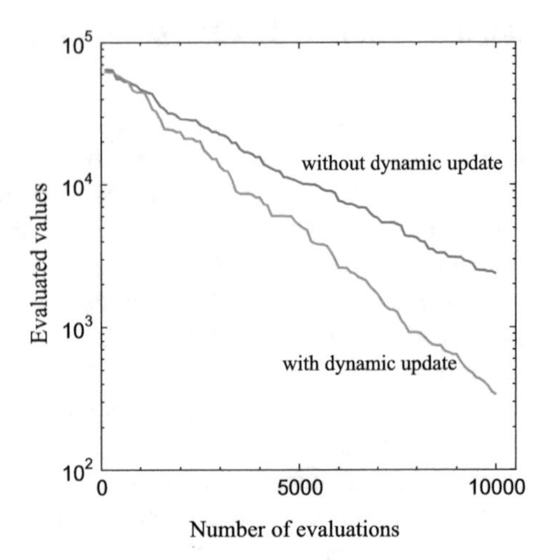

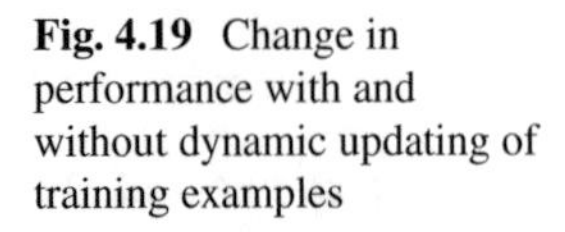

Fig. 4.19 Change in performance with and without dynamic updating of training examples

Training examples are prepared by conducting a standard search for a few generations in DE. However, in DE, vectors evolve and the population becomes increasingly close to the solution as the number of generations increases. Therefore, a vector with a good evaluated value in an early generation could have a worse evaluated value in later generations. This is because a stricter judgment standard is used near the solution. Cases that would be judged as positive examples in earlier generations may have to be judged as negative examples as a result of vector evolution.

Therefore, dynamic updating of training examples with evolution of the vector population is considered. In other words, old training examples are removed as necessary and new ones are more actively used. The use of such a method greatly enhances performance.

Figure 4.19 compares the performance using benchmark function F_1 (see Table 4.18 for the definition). The classification method is standard k-NN without modification, and dynamic updating of the training set is performed (orange line) or not performed (blue line). The feature vector consists of selected individuals ($\vec{r}_1$, $\vec{r}_2$, and $\vec{r}_3$). The solution convergence speed is faster with dynamic updating, and the solution after the same number of evaluations (10,000) is more than one order of magnitude better.

The number of training examples is a problem in dynamic updating. The amount of stored data and the corresponding score are compared for two benchmark functions (see Table 4.15). This table shows that the results are poor when an insufficient amount of training examples is stored. On the other hand, the evaluated value barely changes when there are 500–2,000 samples. This means that an insufficient amount of data is problematic, but the performance does not change much when there is more than a certain amount of samples. The reason is attributed to the evolution of the vector population in DE. The target of classification also evolves and is generated in places very far from old training data, and removing old training examples does not result in noise. The classification method is k-NN; hence, only new training examples

Table 4.15 Change in evaluated value with number of stored training samples

Bench. function	100 samples	200 samples	500 samples	1,000 samples	2,000 samples
F_1	4.27E+02	3.46E+02	3.13E+02	3.37E+02	3.33E+02
F_{11}	5.47E+00	4.43E+00	4.04E+00	3.97E+00	3.85E+00

Table 4.16 Number of positive and negative examples

Benchmark function	Positive examples	Negative examples	Variance
F_1	96.1	903.9	127.69
F_5	38.6	961.4	92.84
F_9	93.3	906.7	109.41
F_{10}	106.5	893.5	49.65

near the center of the classification targets are used. This shows that storing all new training examples does not cause a problem. However, storing too many training examples increases the calculation cost of k-NN classification. Therefore, the number of training examples is set to 1,000 in subsequent experiments.

Classification in NENDE is a binary classification into positive and negative examples. An even ratio (i.e., 1:1) of positive and negative examples in the training set is generally desirable in this type of binary classification. However, as shown above, such an even ratio does not occur in DE search examples. The following table shows the number of positive and negative examples per 1,000 training examples in some benchmark functions (see Table 4.16). The training examples are accumulated 10 times, and the average is shown. The positive-to-negative example ratio is biased 1:9 or more in all benchmark functions. This bias may result in incorrect classification, and thus, some form of correction to the training examples is desirable.

For this purpose, we shall explain how to modify the k-NN method to make a stricter judgement based on the bias between positive and negative examples. The number of selected individuals to be judged as positive examples (denoted as p_{select}) per generation is predetermined, and p_{select} individuals classified to be closer to positive examples are judged to be positive examples in order of likelihood. Others are judged to be negative examples. Individuals closer to positive examples mean individuals with more positive examples among k training examples. In other words, p_{select} individuals with the largest number of positive examples within k nearest training examples are chosen and judged as positive examples.

As a test case, assume $k = 5$, the number of vectors in population is 50, and $p_{select} = 10$. If there were three vectors with five positive examples among the nearest k training examples, five vectors with four positive examples, ten vectors with three positive examples, and the rest of the vectors had two or less positive examples, then the identified positive examples are vectors with five nearby positive examples (three vectors), vectors with four nearby positive examples (five), and vectors with three nearby positive examples (two). There are ten vectors with three positive examples, but as the vector count exceeds p_{select}, two vectors are randomly selected out of ten vectors with three nearby positive examples.

An important parameter, p_{select}, is defined using the following equation:

$$p_{select} = (N \times PN/TS) \times \beta, \tag{4.61}$$

where N is the population size (the number of training samples selected in one generation), PN is the number of positive examples in the training set, and TS is the total number of training examples. This is meant to conduct selection by taking the bias in training examples into account. PN/TS is the fraction of positive examples among the samples. Thus, if $p_{select} = N \times PN/TS$, the ratio of positive example judgments within the total selected individuals in each generation is the same as the ratio of positive examples among the training examples. Even in biased training examples where the positive-to-negative example ratio is 1:9, setting the amount of positive example judgments in each generation to 1/10 of N results in a trend similar to the bias. However, the classification is low in this setting. Therefore, a correction coefficient β is multiplied further. Although $\beta = 0.5$ is adopted empirically, the value of β may have to be adjusted based on the considered function.

In order to confirm the effectiveness of extending the k-NN method as shown above, the performance of the following three methods is compared using several benchmark functions:

Method 1 Standard k-NN.
Method 2 k-NN with p_{select} (improved k-NN as discussed in the above p_{select})
Method 3 SVM (SVM-Light[5] was used)

Pure SVM does not work well because of the bias in the learning examples. Therefore, the samples are corrected such that the number of positive and negative examples are the same. This is achieved by taking the average of the positive examples and choosing half of all samples as positive examples in order of proximity to the average.

The typical result of the above experiment for F_1 is shown in Fig. 4.20. This result shows that improved k-NN increases the performance of DE, and the SVM demonstrates worse performance than standard k-NN. One possible reason for this is that old training samples are unusable because individuals continue to evolve in DE. SVM is a method that uses all stored training samples; thus, old training samples could exert adverse effects and ultimately result in a significantly bad precision. Therefore, the SVM method that makes global judgments is not suitable in situations such as DE where the classification standards change. In contrast, methods that focus on local portions (including k-NN) show good performance.

In standard classifier learning, the data to be classified are converted by certain rules into feature vectors, which are then mapped one-to-one to a feature space. Classification is then conducted based on positions and distances on the feature space. As

[5]SVM-Light Support Vector Machine, http://svmlight.joachims.org/.

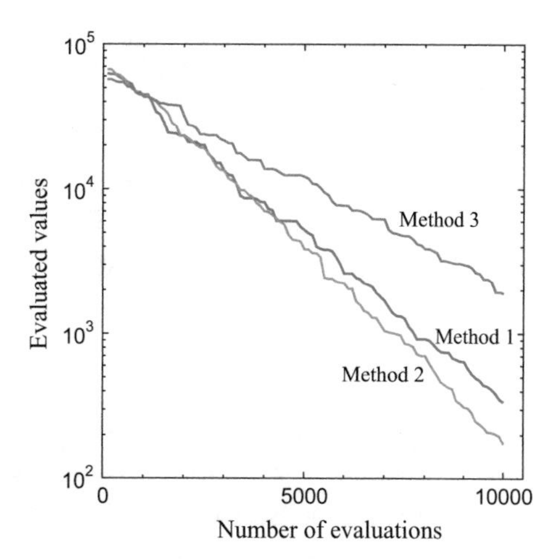

Fig. 4.20 Evaluation of performance of various classification methods

a consequence, how feature vectors are generated from data, i.e., how feature vectors are appropriately defined, significantly affects classification precision. Inappropriate feature vector settings deteriorate classification accuracy; thus, determining which feature vectors to use is an important issue in classifier design.

NENDE classifies individual selection during mutation, where three vectors, $\vec{r}_1, \vec{r}_2, \vec{r}_3$, must be chosen. These three vectors may be individually classified. However, as only one mutation vector is ultimately generated, combining three vectors $(\vec{r}_1, \vec{r}_2, \vec{r}_3)$ into one vector would be appropriate. Therefore, the feature vector is some combination of three vectors $(\vec{r}_1, \vec{r}_2, \vec{r}_3)$. The objective of classification is to conduct selection of superior individuals, i.e., generation of a mutation vector superior to the parent vector, even though the classification target is $\vec{r}_1, \vec{r}_2, \vec{r}_3$. Therefore, a useful feature vector may be generated by using mutation vectors other than $\vec{r}_1, \vec{r}_2, \vec{r}_3$. The performance of some feature vectors is compared and discussed below.

Features that are considered useful in this classification are as follows:

- Selected vectors $\vec{r}_1, \vec{r}_2, \vec{r}_3$,
- Mutation vector $\vec{v}$, see Eq. (2.5),
- Parent vector $\vec{x}$.

Performance is compared for the five features defined below (denoted as $\vec{fv}_1$–$\vec{fv}_5$):

1. $\vec{fv}_1 = (\vec{r}_1, \vec{r}_2, \vec{r}_3)$
   Selected individuals are concatenated as a feature vector. The dimension of this vector is three times the dimension of each individual vector $(\vec{r}_1, \vec{r}_2, \vec{r}_3)$.
2. $\vec{fv}_2 = \vec{v}$
   The mutation vector, $\vec{v}$, is used as the feature vector.
3. $\vec{fv}_3 = (\vec{v}, \vec{x})$
   The mutation vector, $\vec{v}$, and the parent vector, $\vec{x}$, are concatenated as a feature vector.

Table 4.17 Performance comparison for various feature vectors

Benchmark function	$\vec{fv}_1$	$\vec{fv}_2$	$\vec{fv}_3$	$\vec{fv}_4$	$\vec{fv}_5$
F_1	2.16E+02	**1.85E+01**	1.42E+03	3.83E+01	4.33E+02
F_4	4.41E+04	**3.09E+04**	7.73E+04	**3.08E+04**	5.13E+04
F_6	2.49E+06	**2.11E+05**	3.76E+07	7.72E+05	8.32E+06
F_9	2.32E+02	**2.17E+02**	2.57E+02	**2.06E+02**	2.32E+02

4. $\vec{fv}_4 = (\vec{r}_1, \vec{r}_2 - \vec{r}_3)$
   The individual vector, $\vec{r}_1$, and difference vector $(\vec{r}_2 - \vec{r}_3)$ are concatenated as a feature vector. $\vec{r}_1, \vec{r}_2, \vec{r}_3$ are often not treated equally in mutation in DE, but instead are used as an individual vector, $\vec{r}_1$, and a difference vector $(\vec{r}_2 - \vec{r}_3)$ as shown in Eq. (2.5); hence, this feature vector was adopted.
5. $\vec{fv}_5 = (\vec{r}_1, \vec{r}_2 - \vec{r}_3, \vec{x})$
   $\vec{fv}_4$ and the parent vector are concatenated in this feature vector.

Performance was compared using the above five features using benchmark functions (see Table 4.18 for the definition). Table 4.17 shows the average of 10 runs. The values in bold show the best performance (two bold values indicate no significant difference within 5% significance level of a t-test). This result shows that simply using the mutation vector, $\vec{v}$, as the feature vector yields the best results in almost all functions, followed by the feature vector concatenating $\vec{r}_1$ and the difference vector $(\vec{r}_2 - \vec{r}_3)$. The results from vectors including information of the parent vector $\vec{x}$, which are $\vec{fv}_3$ and $\vec{fv}_5$, are not very good.

Why does the performance worsen when parent vector information is included? When focusing on only one vector, evolution proceeds or does not proceed based on the results of the comparison of parent and child vectors. However, the entire vector population gradually moves toward the solution in DE. The child vector to be generated should be superior in the population and to the parent. If the selected parent vector is not superior in the population, the child might be inferior in the population even if it is superior to the parent. In other words, adding information on the parent vector makes the classification standard less stringent.

The claim that using the mutation vector directly as the feature vector resulted in the best performance is still open for discussion. The mutation vector is directly linked to the child vector; thus, examining the mutation vector during classification has a large effect. Therefore, $\vec{fv}_2$ is adopted as the feature vector in NENDE, i.e., the mutation vector is used as the feature vector as it is.

NENDE can be summarized as follows:

- Accumulation of training examples
  DE is performed for a certain number of generations to accumulate training examples. The accumulated results will continue to be used to dynamically update the training examples. Old training examples are discarded in order once 1,000 training examples are stored.

- Classification method
  Extended k-NN is used. In each generation, the p_{select} examples (as defined in Eq. (4.61)) are classified as positive examples in order of proximity to the positive examples.
- Selection of feature vectors
  The mutation vector, $\vec{v}$, is used as the feature vector.

The actual NENDE process can be summarized as shown below (see also Algorithm 4.5):

Step1 Perform DE for a certain number of generations to accumulate training examples for the classifier.

Step2 Individuals ($\vec{r}_1$, $\vec{r}_2$, and $\vec{r}_3$) are randomly selected for each of the vectors $\vec{x}_i$ ($i = 1, \ldots, N$).

Step3 Take the mutation vector, $\vec{u}_i$, obtained from $\vec{r}_1$, $\vec{r}_1$ and $\vec{r}_3$ as the feature vector of this individual.

Step4 For each of the N feature vectors, take k training examples that are the closest neighbors and count the number of positive examples (0 to k).

Step5 Select p_{select} out of N feature vectors that have a larger number of positive examples among training examples, and classify these as positive examples. The rest are classified as negative examples.

Step6 Proceed to crossover and subsequent processes for the sets of individuals ($\vec{r}_1$, $\vec{r}_2$, and $\vec{r}_3$) classified as positive examples in **Step5**. The parent vector remains unchanged in the next generation for the negative examples.

Step7 For those that are subject to crossover and subsequent processes (i.e., those classified as positive in **Step5**), both the feature vector and the class obtained from the results in the selection phase are added as a training example (positive example if evolved, negative example otherwise).

Step8 **Steps** from 2 to 7 are repeated until the evaluated value from the evaluation function becomes a certain value. Old training examples are removed in historical order once 1,000 training examples are accumulated.

Performance was evaluated by comparing NENDE to the methods given below, by using 24 benchmark functions.

1. DE: Standard DE,
2. ILSDE (Sect. 4.6.1) with $K = 3$ (see Eq. (4.57)),
3. NENDE with the following limitation:
   - Categorization method: standard k-NN ($k = 5$),
   - Training samples: fixed once initially determined,
   - Feature vector: $(\vec{r}_1, \vec{r}_2, \vec{r}_3)$,
4. NENDE
   - Categorization method: extended k-NN ($k = 5$),
   - Training samples: dynamically updated,
   - Feature vector: $\vec{v}$ (i.e.,mutation vector).

Algorithm 4.5 NENDE algorithm

```
1: Initialize(x⃗)
2: for g = 1 to G do                                        ▷ Collect training data
3:     for i = 1 to N do
4:         Select r⃗1, r⃗2, r⃗3
5:         v⃗[i] = r⃗1 + F × (r⃗2 − r⃗3)
6:         f⃗v = (x⃗[[i]; v⃗[i]; r⃗1, r⃗2, r⃗3, ...)                 ▷ f⃗v: feature vector
7:         do Crossover                     ▷ u⃗[i] generated by DE process, Eq. (2.6)
8:         x⃗[i] =Selection(x⃗[i], u⃗[i])
9:         if u⃗[i] is selected then
10:            class = +1
11:        else
12:            class = −1
13:        end if
14:        Archive (f⃗v) into DB
15:    end for
16: end for
17: while (evaluation count < Emax) do                      ▷ DE uses classifier
18:    for i = 1 to N do
19:        Select r⃗1, r⃗2, r⃗3
20:        v⃗[i] = r⃗1 + F × (r⃗2 − r⃗3)
21:        f⃗v[i] = (x⃗[[i]; v⃗[i]; r⃗1, r⃗2, r⃗3, ...)              ▷ f⃗v: feature vector
22:        pc[i] = k-neighbor(f⃗v[i], DB)                    ▷ pc: positive count
23:    end for
24:    Sort f⃗v[1], f⃗v[1], ..., f⃗v[N]   ▷ in a descending order of positive counts pc[i]
25:    for i = 1 to p_select do
26:        cp =pop from sorted f⃗v
27:        do Mutation–Crossover–Selection   ▷ u⃗[i] generated by DE process, Eq. (2.6)
28:        if (u⃗[i] is selected) then
29:            class = +1
30:        else
31:            class = −1
32:        end if
33:        Archive (cp) into DB
34:    end for
35: end while
```

For the sake of comparison, we set the number of function evaluations to be 10,000 for all methods, and the results were averaged over 10 runs. As for the DE settings, we used vector population $N = 100$, weight of the difference vector $F = 0.8$ (Eq. (2.5)), and crossover rate $CR = 0.9$ (used in Eq. (2.6)). In case of NENDE, we set the number of stored training samples to be 1,000 and the number of generations where standard DE is employed to accumulate training samples G to be 10.

Two sets of benchmark functions were selected for performance evaluation. The first set is 15 functions obtained from reference [36], i.e., F_1–F_{15}. The second set is nine functions obtained from reference [31], i.e., FF_1–FF_{10}, where FF_5 is excluded. Tables 4.18 and 4.19 show all these functions (see also Figs. 4.21, 4.22 and 4.23). In the equations in Table 4.19, $bias$ is a constant to shift the value of the global optimum, o is the value to be shifted, and M is a rotation matrix for rotation. The minimum

Table 4.18 Benchmark functions from [36]

	Function name	Definition	Ranges
F_1	Sphere	$\sum_{i=1}^{30} x_i^2$	$[-100, 100]^{30}$
F_2	Schwefel's Problem 2.22	$\sum_{i=1}^{30} \mid x_i \mid + \prod_{i=1}^{30} \mid x_i \mid$	$[-10, 10]^{30}$
F_3	Schwefel's Problem 1.2	$\sum_{i=1}^{30} \left(\sum_{j=1}^{i} x_j\right)^2$	$[-100, 100]^{30}$
F_4	Schwefel's Problem 2.21	$\max_{1 \le i \le 30} \mid x_i \mid$	$[-100, 100]^{30}$
F_5	Rosenbrock	$\sum_{i=1}^{29} [100(x_{i+1} - x_i^2)^2 + (x_i - 1)^2]$	$[-30, 30]^{30}$
F_6	Step	$\sum_{i=1}^{30} (\lfloor x_i + 0.5 \rfloor)^2$ The floor function, $\lfloor x \rfloor$, gives the largest integer less than or equal to x	$[-100, 100]^{30}$
F_7	Quartic (+ Noise)	$\sum_{i=1}^{30} i x_i^4 + \text{random}[0, 1)$	$[-1.28, 1.28]^{30}$
F_8	Schwefel's Problem 2.26	$-\sum_{i=1}^{30} (x_i \sin(\sqrt{\mid x_i \mid}))$	$[-500, 500]^{30}$
F_9	Rastrigin	$\sum_{i=1}^{30} [x_i^2 - 10\cos(2\pi x_i) + 10]$	$[-5.12, 5.12]^{30}$
F_{10}	Ackley	$-20\exp(-0.2\sqrt{\frac{1}{30} \sum_{i=1}^{30} x_i^2}) - \exp(\frac{1}{30} \sum_{i=1}^{30} \cos 2\pi x_i)$ $+20 + e$	$[-32, 32]^{30}$
F_{11}	Griewank	$\frac{1}{4000} \sum_{i=1}^{30} x_i^2 - \prod_{i=1}^{30} \cos(\frac{x_i}{\sqrt{i}}) + 1$	$[-600, 600]^{30}$
F_{12}	Penalized(1)	$\frac{\pi}{30}\{10\sin^2(\pi y_1) + \sum_{i=1}^{29} (y_i - 1)^2[1 + 10\sin^2(\pi y_{i+1})] + (y_{30} - 1)^2\} + \sum_{i=1}^{30} u(x_i, 10, 100, 4)$ $y_i = 1 + \frac{1}{4}(x_i + 1)$ $u(x_i, a, k, m) = \begin{cases} k(x_i - a)^m, & x_i > a \\ 0, & -a \le x_i \le a \\ k(-x_i - a)^m, & x_i < a \end{cases}$	$[-50, 50]^{30}$

(continued)

Table 4.18 (continued)

	Function name	Definition	Ranges
F_{13}	Penalized(2)	$0.1\{\sin^2(3\pi x_1) + \sum_{i=1}^{29}(x_i - 1)^2[1 + \sin^2(3\pi x_{i+1})]$ $+(x_{30} - 1)^2[1 + \sin^2(2\pi x_{30})]\} + \sum_{i=1}^{30} u(x_i, 5, 100, 4)$	$[-50, 50]^{30}$
F_{14}	Shekel's Foxholes	$\left[\frac{1}{500} + \sum_{j=1}^{25} \frac{1}{j+\sum_{i=1}^{2}(x_i - a_{ij})^6}\right]^{-1}$ $a_{ij} = \begin{pmatrix} -32 & -16 & 0 & 16 & 32 & -32 & \cdots & 0 & 16 & 32 \\ -32 & -32 & -32 & -32 & -32 & -16 & \cdots & 32 & 32 & 32 \end{pmatrix}$	$[-65.536, 65.536]^2$
F_{15}	Kowalik	$\sum_{i=1}^{11}\left[a_i - \frac{x_1(b_i^2 + b_i x_2)}{b_i^2 + b_i x_3 + x_4}\right]^2$ $a_i = \{0.1957, 0.1947, 0.1735, 0.1600, 0.0844, 0.0627, 0.0456, 0.0342, 0.0323, 0.0235, 0.0246\}$ $b_i^{-1} = \{0.25, 0.5, 1, 2, 4, 6, 8, 10, 12, 14, 16\}$	$[-5, 5]^4$

Table 4.19 Benchmark functions from [31]

	Function name	Definition	Ranges
FF_1	Shifted sphere	$\sum_{i=1}^{30} z_i^2 + bias, \quad z = x - o$	$[-100, 100]^{30}$
FF_2	Shifted Schwefel's Problem 1.2	$\sum_{i=1}^{30} (\sum_{j=1}^{i} z_j)^2 + bias, \quad z = x - o$	$[-100, 100]^{30}$
FF_3	Shifted rotated high conditioned eliptic	$\sum_{i=1}^{30} (10^6)^{\frac{i-1}{30-1}} z_i^2 + bias, \quad z = (x - o) \times M$ M: orthogonal matrix	$[-100, 100]^{30}$
FF_4	Shifted Schwefel's Problem 1.2(+Noise)	$\sum_{i=1}^{30} (\sum_{j=1}^{i} z_j)^2 \times (1 + 0.4 \times \mid N(0,1) \mid) + bias,$ $z = x - o$	$[-100, 100]^{30}$
FF_6	Shifted Rosenbrock	$\sum_{i=1}^{30-1} [100(z_{i+1} - z_i^2)^2 + (z_i - 1)^2] + bias,$ $z = x - o + 1$	$[-100, 100]^{30}$
FF_7	Shifted rotated Griewank (without bounds)	$\frac{1}{4000} \sum_{i=1}^{30} z_i^2 - \prod_{i=1}^{30} \cos(\frac{z_i}{\sqrt{i}}) + 1 + bias,$ $z = (x - o) \times M$ M:linear transformation matrix	$[0, 600]^{30}$
FF_8	Shifted rotated Ackley (global optimum on bounds)	$-20\exp(-0.2\sqrt{\frac{1}{n} \sum_{i=1}^{30} z_i^2}) - \exp(\frac{1}{30} \sum_{i=1}^{30} \cos 2\pi z_i)$ $+20 + e + bias, z = (x - o) \times M$ M:linear transformation matrix	$[-32, 32]^{30}$
FF_9	Shifted Rastrigin	$\sum_{i=1}^{30} [z_i^2 - 10\cos(2\pi z_i) + 10] + bias, \quad z = x - o$	$[-5, 5]^{30}$
FF_{10}	Shifted rotated Rastrigin	$\sum_{i=1}^{30} [z_i^2 - 10\cos(2\pi z_i) + 10] + bias,$ $z = (x - o) \times M$ M:linear transformation matrix	$[-5, 5]^{30}$

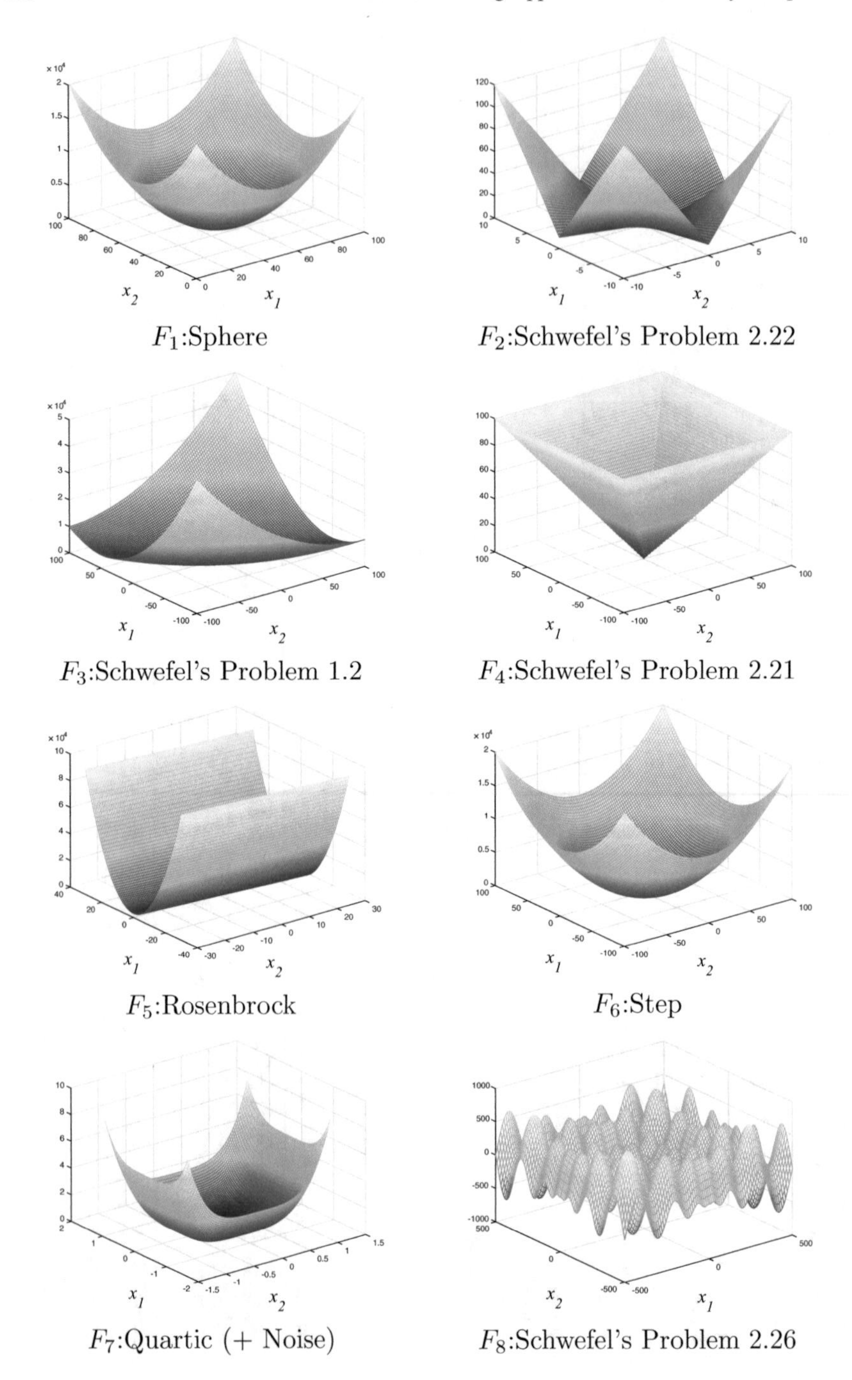

Fig. 4.21 Benchmark functions from [36] (1)

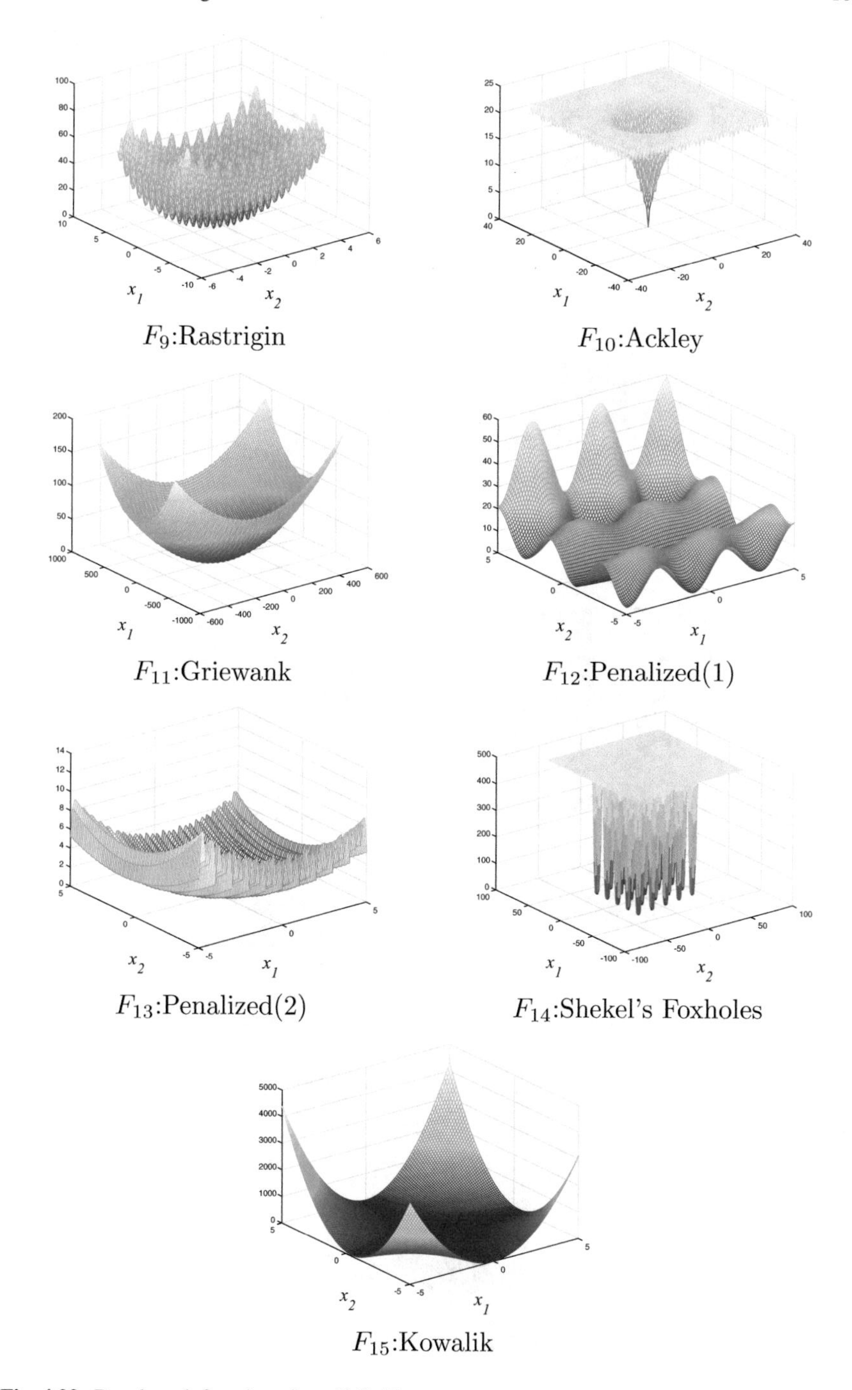

Fig. 4.22 Benchmark functions from [36] (2)

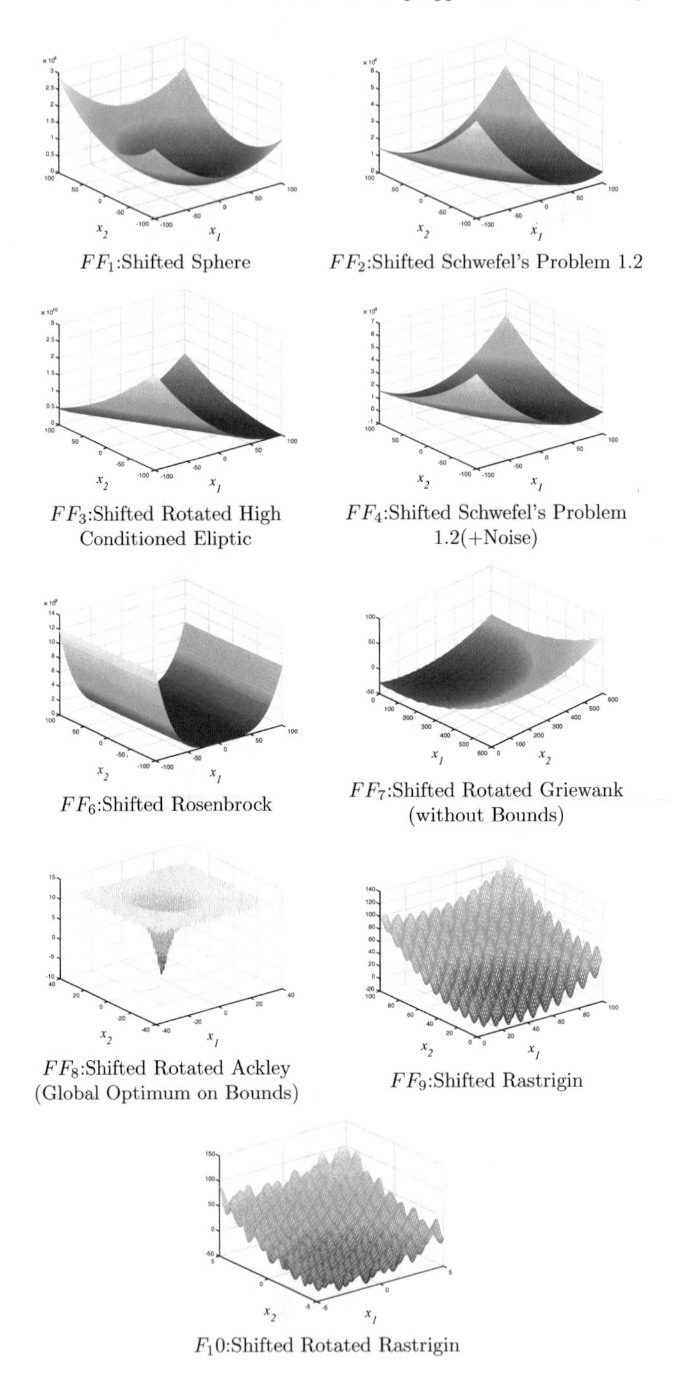

FF_1:Shifted Sphere

FF_2:Shifted Schwefel's Problem 1.2

FF_3:Shifted Rotated High Conditioned Eliptic

FF_4:Shifted Schwefel's Problem 1.2(+Noise)

FF_6:Shifted Rosenbrock

FF_7:Shifted Rotated Griewank (without Bounds)

FF_8:Shifted Rotated Ackley (Global Optimum on Bounds)

FF_9:Shifted Rastrigin

$F_1 0$:Shifted Rotated Rastrigin

Fig. 4.23 Benchmark functions [31]

Table 4.20 Minimum values for benchmark functions from [36]

	Function name	Optimum
F_1	Sphere	$F_1(0,\ldots,0)=0$
F_2	Schwefel's Problem 2.22	$F_2(0,\ldots,0)=0$
F_3	Schwefel's Problem 1.2	$F_3(0,\ldots,0)=0$
F_4	Schwefel's Problem 2.21	$F_4(0,\ldots,0)=0$
F_5	Rosenbrock	$F_5(1,\ldots,1)=0$
F_6	Step	$F_6(0,\ldots,0)=0$
F_7	Quartic (+ Noise)	$F_7(0,\ldots,0)=0$
F_8	Schwefel's Problem 2.26	$F_8(420.9678,\ldots,420.9678)=-12569.5$
F_9	Rastrigin	$F_9(0,\ldots,0)=0$
F_{10}	Ackley	$F_{10}(0,\ldots,0)=0$
F_{11}	Griewank	$F_{11}(0,\ldots,0)=0$
F_{12}	Penalized(1)	$F_{12}(1,\ldots,1)=0$
F_{13}	Penalized(2)	$F_{13}(1,\ldots,1)=0$
F_{14}	Shekel's Foxholes	$F_{14}(-32,-32)\approx 1$
F_{15}	Kowalik	$F_{15}(0.1928, 0.1908, 0.1231, 0.1358)\approx 0.0003075$

values for benchmark functions in Table 4.18 are shown in Table 4.20, whereas the minimum values for benchmark functions in Tables 4.19 are $FF_i(o) = bias$ for all i. The reader is referred to references [31, 36] for more details.

Tables 4.21 and 4.22 show the average and variance of 10 experiments in each function, respectively. For all functions, the goal is to obtain the minimum value, and a smaller value in the result means that the performance is better. The minimum value for the first set of benchmark functions was −1.26E+04 in F_8, 1 in F_{14}, 3.07E-04 in F_{15}, and 0 otherwise. The minimum value for the second set (i.e., FF_1–FF_{10}) is $bias$ (a value added to prevent 0 from becoming the minimum value). Although this bias was included in calculations in the solution search process, the minimum value is corrected to 0 by subtracting $bias$ in the result output.

The best result of the four methods is highlighted in bold. Searches are designed not to exceed the range of initial values set in Tables 4.21 and 4.22. However, this constraint is not imposed on FF_7, because the global optima is outside the bounds of the initial values. Note the following points are clearly shown in the tables.

- The result for F_4 was not obtained using the conventional method.
- No significant difference between ILSDE and NENDE according to a two-sided t-test (significance level 5%) in F_2, F_3, F_4, and F_9.

Figures 4.24 and 4.25 plot the search results of each function. The methods in each graph are indicated as follows:

Table 4.21 Results for benchmark functions from [36] (left: average; right: variance)

Function	DE		ILSDE		Limited NENDE		NENDE	
F_1	3.46E+03	2.99E+05	3.18E+02	1.10E+04	2.70E+03	3.20E+05	**1.15E+01**	1.50E+01
F_2	2.46E+01	8.77E+00	8.28E+00	1.61E+00	2.15E+01	4.10E+00	**1.50E+00**	3.79E-02
F_3	3.99E+04	3.14E+07	3.87E+03	7.63E+05	2.83E+04	3.83E+07	**3.13E+02**	2.89E+04
F_4	6.90E+01	2.53E+01	5.19E+01	1.75E+02	NA	NA	**8.37E+00**	3.45E+00
F_5	1.73E+06	1.20E+12	3.32E+04	4.70E+08	5.75E+05	5.81E+10	**6.65E+02**	1.93E+05
F_6	3.43E+03	6.18E+05	3.51E+02	7.64E+03	2.16E+03	3.49E+05	**1.23E+01**	1.02E+01
F_7	1.05E+00	9.43E-02	1.80E-01	4.12E-03	6.89E-01	3.45E-02	**6.95E-02**	3.54E-04
F_8	−4.43E+03	5.00E+04	−4.71E+03	7.03E+04	−4.66E+03	2.23E+05	**−6.19E+03**	7.02E+05
F_9	2.60E+02	1.11E+02	2.43E+02	3.84E+02	2.52E+02	2.52E+02	**2.14E+02**	1.41E+02
F_{10}	1.29E+01	8.01E-01	5.78E+00	3.39E-01	1.13E+01	6.43E-01	**2.50E+00**	1.65E-01
F_{11}	3.07E+01	1.54E+01	4.10E+00	7.92E-01	2.22E+01	4.37E+01	**1.09E+00**	4.21E-04
F_{12}	2.07E+05	1.84E+10	1.80E+01	4.83E+01	2.12E+04	1.28E+09	**3.49E-01**	1.79E-01
F_{13}	2.31E+06	2.90E+12	3.35E+03	2.37E+07	8.02E+05	1.43E+12	**1.36E+00**	4.08E-01
F_{14}	**1.00E+00**	0.00E+00	**1.00E+00**	0.00E+00	1.20E+00	1.60E-01	1.20E+00	3.59E-01
F_{15}	**4.37E-04**	7.13E-08	5.13E-04	1.30E-07	7.08E-04	1.56E-07	8.92E-04	5.25E-08

Table 4.22 Results for benchmark functions from [31] (left: average; right: variance)

Function	DE		ILSDE		Limited NENDE		NENDE	
FF_1	2.22E+03	2.30E+05	2.14E+02	1.52E+03	2.15E+03	1.99E+05	**1.67E+01**	1.68E+01
FF_2	2.38E+04	2.47E+07	1.07E+04	6.44E+06	2.64E+04	2.77E+07	9.38E+03	2.83E+06
FF_3	1.31E+08	1.27E+15	3.86E+07	1.95E+14	1.21E+08	1.92E+15	4.76E+07	3.29E+14
FF_4	3.46E+04	3.00E+07	1.87E+04	3.00E+07	3.40E+04	2.51E+07	1.41E+04	2.78E+07
FF_6	1.13E+08	4.34E+15	1.39E+06	1.18E+12	4.66E+07	8.01E+14	**7.27E+04**	9.56E+08
FF_7	6.73E+02	9.04E+03	7.26E+01	6.21E+02	6.50E+02	2.37E+04	**2.90E+01**	7.55E+01
FF_8	2.11E+01	5.31E-03	2.11E+01	4.20E-03	2.11E+01	2.38E-03	2.11E+01	6.56E-03
FF_9	2.31E+02	3.01E+02	2.16E+02	1.39E+02	2.26E+02	1.77E+02	**2.04E+02**	1.29E+02
FF_{10}	2.75E+02	3.11E+02	2.39E+02	1.44E+02	2.67E+02	2.78E+02	**2.24E+02**	1.51E+02

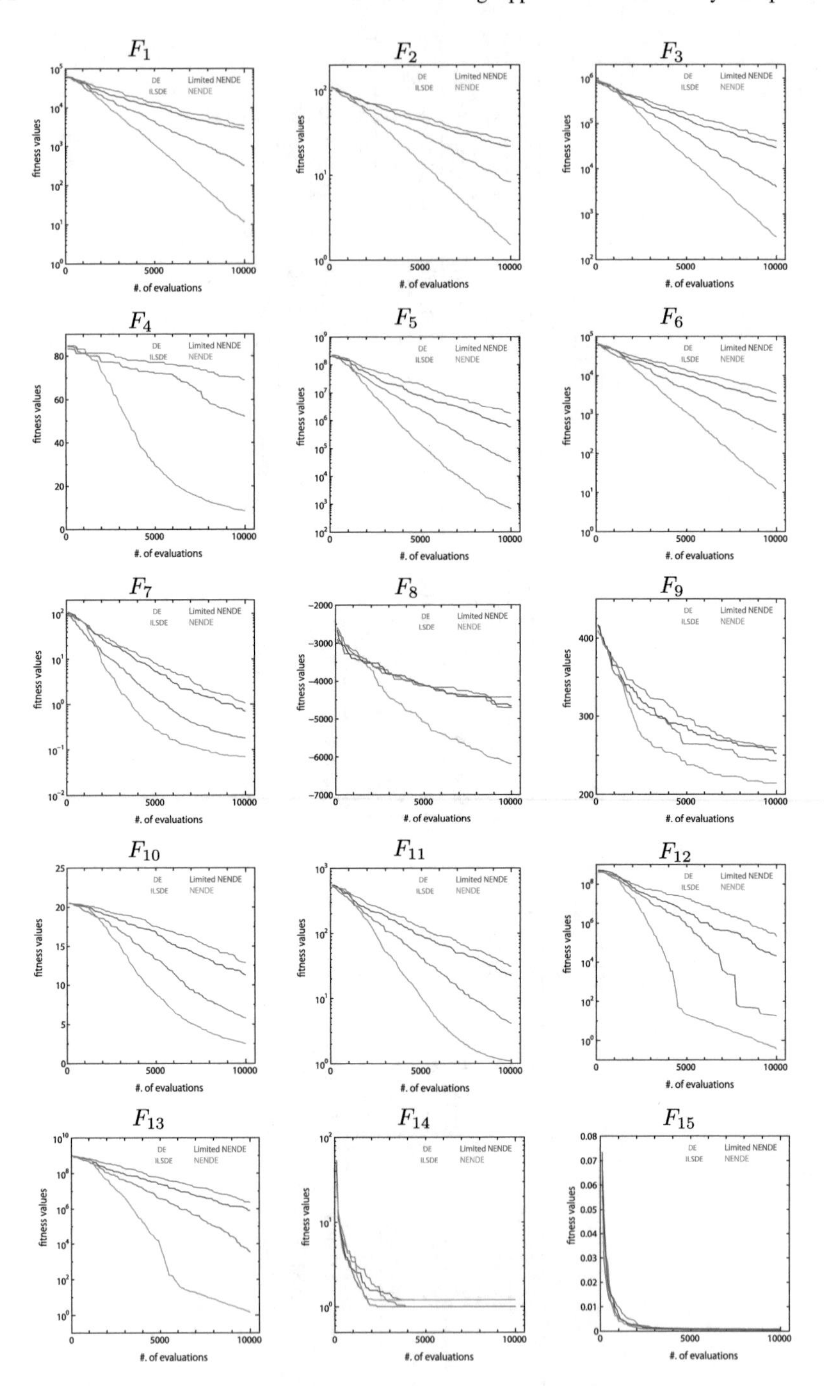

Fig. 4.24 Results for benchmark functions from [36]

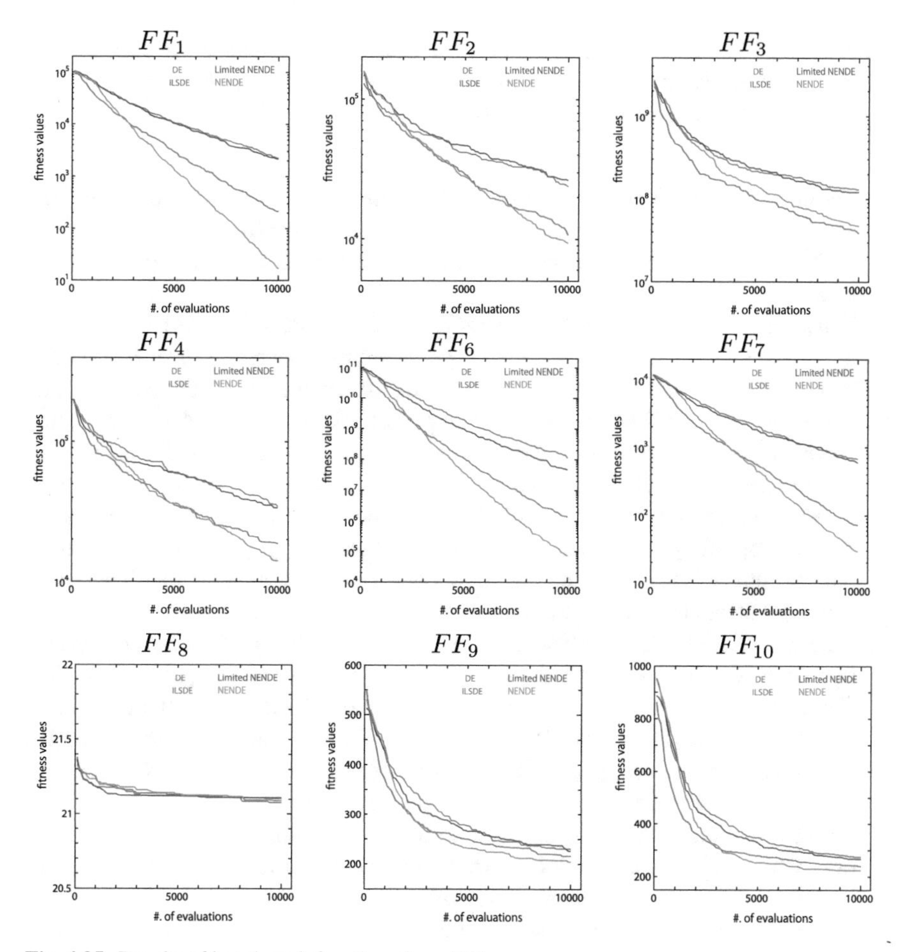

Fig. 4.25 Results of benchmark functions from [31]

- Blue line: DE
- Green line: ILSDE
- Red line: Limited NENDE
- Orange line: NENDE.

In all cases, the horizontal axis represents the number of function evaluations and the vertical axis represents the evaluated value (average of 10 runs). The vertical axis is a linear scale when numbers are shown and a logarithmic scale when powers are shown. Hence, F_4, F_8, F_9, F_{10}, and F_{15}, as well as FF_8, FF_9, and FF_{10}, are linear scale and the rest are logarithmic scale.

Tables 4.21 and 4.22 show that the order of search performance, from good to bad, is NENDE, ILSDE, limited NENDE, and DE in most of the functions. The conventional method is slightly better than DE, but the performance was significantly enhanced after making the improvement and became better than ILSDE.

Each function is analyzed in detail below:

- F_1–F_7
  These functions are pure single-peak functions with no shifts or rotations. The results show that the graphs have almost the same shape except for F_4 and F_7. NENDE always yields the best result, followed by ILSDE, the conventional method, and DE.
  Function F_7 contains noise, and hence, NENDE was proven to work even if there is noise. However, the slope of the graph becomes moderate at over 6,000 evaluations. The reason is attributed to the premature convergence; the current value is 0.1, which is sufficiently close to the global optimum of 0. Ideally, the value should approach 0 more closely from this situation, and NENDE still has room for improvement in this regard. In other words, appropriate evolution is not being conducted once a value close to the solution is attained. One reason may be because of an inappropriately selected p_{select} value. The value of p_{select} is kept constant even if the number of improving individuals per generation becomes smaller. Therefore, evolution stagnates for some cases.
  A solution for F_4 was not obtained using the conventional method because evolution stopped at a point where the classified results all became negative examples. This is caused by bias in the training examples and the selection of standard k-NN as the classification method. Improved NENDE takes bias in training examples into consideration during classification and is designed to always classify a certain number of positive examples; hence, this problem did not appear.
  NENDE demonstrates vastly superior performance to other methods in this function, which has a simple structure where the maximum value is obtained among the value in each dimension of the vector.
  Therefore, superior performance is attained when the value is a similarly small value in all dimensions. NENDE excels in this type of functions because machine learning is effective when the structure of the solution is simple. The value of all dimensions is the same in the global optima for not only F_4, but for all functions in Table 4.18 (with the exception of F_{15}). As a consequence, good performance is expected from NENDE.
- F_8–F_{13}
  These functions are pure multipeak functions with no shifts or rotations. The results are plotted in Fig. 4.24, and NENDE was found to perform best in all cases. In F_8, ILSDE is only slightly better than DE, while NENDE demonstrates vastly superior performance to DE. NENDE is found to perform better than ILSDE in multipeak functions as well.
  The search performance of NENDE for F_{12} and F_{13} decreases at approximately 5,000 evaluations and at evaluated value less than 100. The slope is almost the same as DE after that point. This means that the convergence speed is no longer different from DE. One cause is an inappropriate p_{select} value that led to the classification of many positive examples that should be classified as negative examples, as in

the above case of F_7. Furthermore, bias in the population may be another reason. Function F_9 (Rastrigin function) is a function with many peaks and valleys, as shown in the figure, and DE is a search method that is prone to convergence to a local optimum. Therefore, this function is difficult for DE. Only a few out of 100 individuals evolve per generation beyond a certain point with standard DE. NENDE shows the best performance in this type of function, but the solution search speed slightly slows down at some point.

- F_{14}–F_{15}
These functions are multipeak functions with no shifts or rotations as in the case of F_8–F_{13}, but the vector dimension is smaller. The number of dimensions is 2 in F_{14} and 4 in F_{15}. As shown in Fig. 4.24, reaching the global optimum is simple because of the small number of dimensions, and the global optimum is reached after approximately 2,000–3,000 function evaluations in all methods. There is no large difference in convergence speed, although DE is slightly worse than others. However, NENDE converges before the global optimum in these two functions, and thus, the results of NENDE are slightly worse than DE. The cause of this phenomenon is the bias in individuals, which occurs very close to the optimum solution, as in the above case of F_7.
- FF_1–FF_4
These functions are single-peak functions with shifts and/or rotations. Figure 4.25 shows that NENDE yields the best results, but there is not much difference in FF_2–FF_4. The average value of NENDE is slightly better than ILSDE in FF_2 and FF_4 but is slightly worse in FF_3, and these are not significant differences in t-tests with significance level 5%. NENDE results are worse than similar functions in F_1–F_{15}; hence, NENDE is slightly weaker in functions with shifts and rotations. When the objective function is shifted, the performance of the conventional method becomes as bad as DE, but NENDE still has performance that is comparable or better than ILSDE.
- FF_6–FF_{10}
These functions are multipeak functions with shifts and/or rotations. NENDE provides the best results for FF_6 and FF_7. The convergence speed of NENDE in multipeak functions is faster than ILSDE and therefore has high performance.
FF_8, FF_9, and FF_{10} have many local optima and therefore are difficult functions to solve with DE. In particular, all methods provided results similar to those of DE in FF_8. The minimum value of this function is zero (when bias is excluded). Convergence at a value around 21 indicates trapping at a local optimum.
FF_9 is a shifted Rastrigin function, and FF_{10} is a shifted and rotated Rastrigin function. These are also functions that are difficult for DE to process. NENDE again gives best results. However, trapping to a local optimum with a value around 200 often occurs although the minimum value is 0. Although NENDE improves results for functions that DE is not good at including FF_9, there is still much that can be improved.

The above results indicate that NENDE demonstrates higher overall performance than ILSDE.

Table 4.23 Comparative results with SVC-DE (left: average; right: variance)

Function	DE		NENDE		SVC-DE	
FF_1	2.22E+03	2.30E+05	1.67E+01	1.68E+01	1.07E-01	1.35E-03
FF_6	1.13E+08	4.34E+15	7.27E+04	9.56E+08	2.54E+03	9.67E+06
FF_8	2.11E+01	5.31E-03	2.11E+01	6.56E-03	2.08E+01	4.37E-03
FF_9	2.31E+02	3.01E+02	2.04E+02	1.29E+02	2.09E+02	1.71E+02
FF_{10}	2.75E+02	3.11E+02	2.24E+02	1.51E+02	2.15E+02	1.88E+02

Next, a comparison of SVC-DE and NENDE is performed. Four (relatively complicated) benchmark functions are used, i.e., FF_1, FF_8, FF_9, FF_{10}. The results obtained after 10,000 function evaluations are presented in Table 4.23. The prediction using SVC-DE is especially effective when the function is simple. For instance, the evaluated value in SCV-DE is one or more orders of magnitude better in the simplest function, FF_1, and in a simple multipeak function, FF_6. On the other hand, there is almost no significant difference between SVC-DE and NENDE in the complicated functions FF_8–FF_{10}. This means that the effectiveness of good individual selection in NENDE becomes evident and results in performance enhancement as the complexity of the function increases.

References

1. Baker Jr., G.A.: The theory and application of the Padé approximant method. Adv. Theor. Phys. **1**, 1 (1965)
2. Bishop, C.M.: Pattern Recognition and Machine Learning. Springer, Berlin (2006)
3. Bollegala, D., Noman, N., Iba, H.: RankDE: learning a ranking function for information retrieval using differential evolution. In: GECCO 11 Proceedings of the 13th Annual Conference on Genetic and Evolutionary Computation, pp. 1771–1778. ACM Press (2011)
4. Brest, J., Greiner, S., Bošković, B., Mernik, M., Žumer, V.: Self-adapting control parameters in differential evolution: a comparative study on numerical benchmark problems. IEEE Trans. Evol. Comput. **10**(6), 646–657 (2006)
5. Cox, D., Little, J., O'shea, D.: Ideals, Varieties, Algorithms, 3rd edn. Springer, Berlin (1992)
6. Dabhi, V.K., Chaudhary, S.: A survey on techniques of improving generalization ability of genetic programming solutions (2012)
7. De Melo, V.V.: Kaizen programming. In: Proceedings of the 2014 an Conference on Genetic and Evolutionary Computation (GECCO), pp. 895–902 (2014)
8. De Melo, V.V.: Breast cancer detection with logistic regression improved by features constructed by Kaizen programming in a hybrid approach. In: Proceedings of the IEEE Congress on Evolutionary Computation (CEC), pp. 16–23 (2016)
9. De Melo, V.V., Banzhaf, W.: Predicting high-performance concrete compressive strength using features constructed by Kaizen programming. In: Proceedings of 2015 Brazilian Conference on Intelligent Systems (BRACIS), pp. 25–30. IEEE (2015)
10. De Melo, V.V., Banzhaf, W.: Improving the prediction of material properties of concrete using Kaizen programming with simulated annealing. Neurocomputing **246**, 25–44 (2017)
11. Dervis, K., Ozturk, C., Karaboga, N., Gorkemli, B.: Artificial bee colony programming for symbolic regression. Inf. Sci. **209**, 1–15 (2012)

12. Drucker, H.: Improving regression using boosting techniques. In: Proceedings of International Conference on Machine Learning (ICML97) (1997)
13. Feng, J.: The relevance vector machine technique for automatic feature selection in genetic programming, Master thesis, Graduate School of Information and Communication Engineering, University of Tokyo (2017)
14. Feng, J., Iba, H.: An evolutionary construction of basis functions based on GP and RVM. In: Proceedings of Evolutionary Computation Symposium, Dec. 10–11, Kujukuri, Chiba, Japan (2016)
15. Gitlow, H., Gitlow, S., Oppenheim, A., Oppenheim, R.: Tools and Methods for the Improvement of Quality. Irwin Series in Quantitative Analysis for Business. Taylor & Francis, New York (1989)
16. Hettiarachchi, D.S., Iba, H.: An evolutionary computational approach to humanoid motion planning. Int. J. Adv. Robot. Syst. **9**(167) (2012)
17. Iba, H.: Bagging, boosting, and bloating in genetic programming. In: Proceedings of Genetic and Evolutionary Computation Conference (GECCO-1999) (1999)
18. Imai, M.: Kaizen (Ky'zen), the Key to Japan's Competitive Success. McGraw-Hill, Singapore (1986)
19. Karaboga, D., Basturk, B.: A powerful and efficient algorithm for numerical function optimization: artificial bee colony (ABC) algorithm. J. Global Optim. **39**(3), 459–471 (2007)
20. Karaboga, D., Ozturk, C., Karaboga, N., Gorkemli, B.: Artificial bee colony programming for symbolic regression. Inf. Sci. **209**, 1–15 (2012)
21. Kera, H., Iba, H.: Vanishing ideal genetic programming. In: Proceedings of the IEEE Congress on Evolutionary Computation (CEC), pp. 5018–5025 (2016)
22. Liu, Y., An, A., Huang, X.: Boosting prediction accuracy on imbalanced datasets with SVM ensembles. Advances in Knowledge Discovery and Data Mining (PAKDD 2006). Lecture Notes in Computer Science, pp. 107–118 (2006)
23. Lu, X., Tang, K., Yao, X.: Classification-assisted differential evolution for computationally expensive problems. In: Proceedings of IEEE Congress on Evolutionary Computation (CEC2011), pp. 1986–1993 (2011)
24. McDermott, J., White, D.R., Luke, S., Manzoni, L., Castelli, M., Vanneschi, L., Jaskowski, W., Krawiec, K., Harper, R., De Jong, K., et al.: Genetic programming needs better benchmarks. In: Proceedings of the 14th Annual Conference on Genetic and Evolutionary Computation, pp. 791–798. ACM (2012)
25. Michael, S., Lipson, H.: Distilling free-form natural laws from experimental data. Science **324**(5923), 81–85 (2009)
26. Möller, H.M., Buchberger, B.: The construction of multivariate polynomials with preassigned zeros. In: Proceedings of the European Computer Algebra Conference on Computer Algebra, pp. 24–31 (1982)
27. Pagie, L., Hogeweg, P.: Evolutionary consequences of coevolving targets. Evol. Comput. **5**(4), 401–418 (1997)
28. Poli, R., Langdon, W.B., McPhee, N.F., Koza, J.R.: A Field Guide to Genetic Programming, Lulu. com (2008)
29. Rad, H.I., Feng, J., Iba, H.: GP-RVM: efficient genetic programming-based symbolic regression using relevance vector machine, submitted to the 2018 IEEE International Conference on Systems, Man, and Cybernetics (SMC) (2018)
30. Storn, R., Price, K.V.: Differential evolution -a simple and efficient heuristic for global optimization over continuous spaces. J. Global Optim. **11**(4), 341–359 (1997)
31. Suganthan, P.N., Hansen, N., Liang, J.J., Deb, K., Chen, Y.P., Auger, A., Tiwari, S.: Problem definitions and evaluation criteria for the CEC 2005 special session on real-parameter optimization. Technical report, vol. 2005005, Nanyang Technological University, Singapore (2005)
32. Suzuki, K., Iba, H.: PSO with clustering by means of affinity propagation. In: Proceedings of Evolutionary Computation Symposium, Dec. 10–11, Kujukuri, Chiba, Japan (2016)

33. Tipping, M.E., Faul, A.C.: Fast marginal likelihood maximization for sparse Bayesian models. In: Proceedings of AISTATS, pp. 1–8 (2003)
34. Vladislavleva, E.J., Smits, G.F., den Hertog, D.: Order of nonlinearity as a complexity measure for models generated by symbolic regression via pareto genetic programming. IEEE Trans. Evol. Comput. **13**(2), 333–349 (2009)
35. Xiaofen, L., Tang, K., Yao, X.: Classification assisted differential evolution for computationally expensive problems. In: Proceedings of the IEEE Congress on Evolutionary Computation (CEC), pp. 1986–1993 (2011)
36. Yao, X., Liu, Y., Lin, G.: Evolutionary programming made faster. IEEE Trans. Evol. Comput. **3**, 82–102 (1999)
37. Yang, X.-S.: Firefly algorithm, stochastic test functions and design optimisation. Int. J. Bio-Inspired Comput. **2**(2), 78–84 (2010)
38. Zhan, Z.-H., Zhang, J., Li, Y., Chung, H.S.-H.: Adaptive particle swarm optimization. IEEE Trans. Syst. Man Cybern. Part B (Cybern.) **39**(6), 1362–1381 (2009)

Chapter 5
Evolutionary Approach to Gene Regulatory Networks

All life is problem solving. I have often said that from the amoeba to Einstein there is only one step.

(Karl Popper)

Abstract Gene regulatory networks (GRNs) described in this chapter are recently attracting attention as a model that can learn in a way similar to neural networks. Gene regulatory networks express the interactions between genes in an organism. We first give several inference methods to GRN. Then, we explain the real-world application of GRN to robot motion learning. We show how GRNs have generated effective motions to specific humanoid tasks. Thereafter, we explain ERNe (Evolving Reaction Network), which produces a type of genetic network suitable for biochemical systems. ERNe's effectiveness is shown by several in silico and in vitro experiments, such as oscillator syntheses, XOR problem solving, and inverted pendulum task.

Keywords DREAM (Dialogue for Reverse Engineering Assessments and Methods) · INTERNe (IEC-based GRN inference with ERNe) · MONGERN (MOtioN generation by GEne regulatory networks) · ERNe (Evolving Reaction Network) · DNA PEN Toolbox · Speciation

5.1 Overview of Gene Regulatory Networks

As a result of the recent revival in Evo-Devo (Evolutionary Developmental Biology[1]), dramatic changes are occurring in regard to our understanding of the genetic control networks of organisms. For example, there have been many surprising discoveries in relation to Hox genes (master control genes specifying segments in the majority

[1]In this field, the process in which organisms occur is researched from an evolutionary perspective. Its purpose is a comprehensive and empirical understanding of how systems and individuals emerge.

H. Iba, *Evolutionary Approach to Machine Learning and Deep Neural Networks*, https://doi.org/10.1007/978-981-13-0200-8_5

of metazoan gates). The Hox genes are stored by a large number of species, and these are linked to diversity in the body plan of animals. Furthermore, the Hox genes are arranged along the chromosomes so that they appear in the same order along the longitudinal axis of the embryo. The mutations in the Hox genes bring about large-scale local changes in the phenotypes. However, there are many factors that are unclear in relation to how selection pressure molds the evolution and diversity of the gene. In parallel with this, neuroscience researchers and evolutionary biologists believe that modularity (integration of functionally linked structures and separation of unrelated structures) is essential on both the phenotype and genotype levels in order to develop complex structures. Furthermore, through the differentiation and overlapping of certain characteristics, it is possible to achieve a functional role different to the copied characteristic through the evolution process. This kind of process is known as exaptation.[2] At the genetic level, this is a similar function to genetic overlapping.

The proteins created by genes affect the activity of other genes. The nodes (genes) on the gene network bond with others to whom they provide activity (linked with arcs and edges). Because of this reason, the bonding between genes on the gene network is indirect. This is a major difference with neural networks. In the case of neural networks, neurons mutually influence each other through direct bonding of synapses.

In the gene network model or GRNs (gene regulatory networks), we should note that genes simply affect growth and development, and that the realization of the final form and function is emergently determined. If genes are activated, gene-specific matter is created, and this contributes to activation and deactivation of the movement of separate genes. In this way, interactions between genes are promoted. Growth and development are affected by the environment. The emergence of individual bodies occurs based on a genetic network in which extremely complex behavior is demonstrated from interactions between genes. Therefore, even in the case of a comparatively small-scale network, it is possible to produce a great amount of complexity. For example, this is suited to realizing the exaptation discussed previously. In other words, the evolution of the genetic network is verified through the emergence of individual bodies and can be used for the newly acquired functions.

The genetic network models the interaction of genes, which are the basis of living phenomena. For this reason, analyzing the genetic network promises to allow us to approach the basic mechanisms of life systems. Important mechanisms for this are robustness and evolvability.

- **Robustness**: Maintain functionality in relation to random mutations, such as genetic transformations due to physical or chemical changes within the environment.
- **Evolvability**: As a result of the genetic transformations, changes occur leading to the acquisition of novel functions and adaptation to the new environment.

[2]This is to utilize a structure that already exists for a new function. For example, it is thought that the feathers of a bird from a dinosaur that evolved with the objective of retaining warmth are used for flying later with their evolution to wings.

These two mechanisms are observed on a wide variety of biological organism levels from genetic control to the fitness of organisms.

The robustness and evolvability of the genetic network are defined as follows.

- **Robustness**: When the gene mutates, the phenotype (actions and functions that occur) is unchanged.
- **Evolvability**: New phenotypes are acquired as a result of genetic mutation, and this contributes to survival.

Stuart Kauffman et al. used a genetic network known as a Random Boolean Graph (RBG) to analyze these mechanisms in detail [10]. With RBG, the input/output relationship of the Boolean function takes a network structure. RBG simply models the control relationships of the expressional relationships of genes (active/non-active). RBG attractors (state of equilibrium/constancy in mechanics) are seen as the destiny (final state) of the cells. This refers to the final state of the genetic expression. A feature of the attractor is that they are a state that is stable in relation to small disturbances, and demonstrate that the network is pulled back to the attractor [1]. On the other hand, there are mutations that cause new attractors. More specifically, the following kinds of network changes are observed due to mutations:

- **Identity**: An original attractor is identical to one of the transformed attractors.
- **Expansion**: One of the original attractors acquires more configurations in the mutated network and preserves all its old ones.
- **Contraction**: One of the original attractors loses some of its configurations in the mutated network.
- **Innovation**: A fully new attractor emerges in the mutated network.

Kauffman et al. empirically verified that if RBG is in a critical state (state positioned between order and chaos[3]) that genetic transformations have robustness and evolvability. In other words, ordered state and critical state networks are essentially robust. Furthermore, in regard to evolvability, the critical state network has a high probability of demonstrating those features. According to this study [1], robustness and evolvability coexist, and this promises to deepen our understanding of the origin of diversity in stable life systems.

The biological process is extremely complex. Normally, for the GRN model, the following two simplified assumptions are required:

- Genetic expression controls are in the regulation of gene transcription.
- The expression of genes and generation of proteins occur consecutively.

In the simple GRN model, a cell consists of one genome and several types of proteins. Genomes are made up of several genes that interact through generated proteins. Each gene has a regulating unit and a structure unit. The regulating unit stipulates particular proteins that suppress or activate this expression. On the other hand, the construction unit describes proteins generated when the genes are expressed. When

[3]This is referred to as the edge of chaos, and is the hypothesis that life has evolved in those regions.

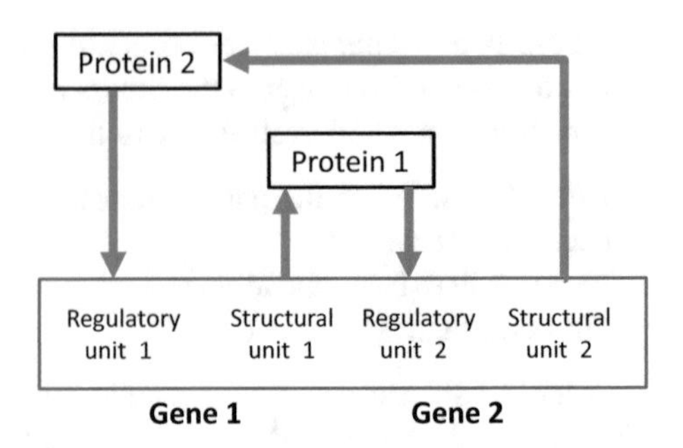

Fig. 5.1 An example of GRN with two genes

the genes are expressed, this means that the expression level is greater than a certain threshold. Figure 5.1 is an example of a two-gene GRN. In this figure, the product of gene 1 suppresses the expression of gene 2, and the product of gene 2 suppresses the expression of gene 1.

5.2 GRN Inference by Evolutionary and Deep Learning Methods

5.2.1 Inferring Genetic Networks

Technologies such as DNA microarrays and oligonucleotide chips have made it possible to observe thousands of genes under a wide variety of conditions (Fig. 5.2). By monitoring transcriptomes[4] on a genome-wide scale, scientists are deepening their understanding of the structure and dynamic changes in the genomic activities at different stages of cellular development.

Inferring genetic networks is a kind of inverse problem, which involves finding the structures of networks with complex interconnections on the basis of the expression levels of each gene. Inverse problems for multivariate dynamical networks are computationally hard. As more parameters are added, the size of the search space and the amount of required data both grow exponentially, reflecting the increased complexity of the problem.

The difficultly of the problem is also due to the characteristics of the data. With current technologies, the noise levels in data from DNA microarray experiments reportedly range from 30% to as high as 200%. Moreover, the time series that can be obtained from a single reaction process is extremely short, with no more than a dozen or so discrete data points captured. The experiments are expensive to perform, and environments where experiments can be repeated easily are yet to be set up. This means that there is no choice but to accept incomplete data or data with varying error levels.

[4]All of the transcripts (mRNA) within a cell.

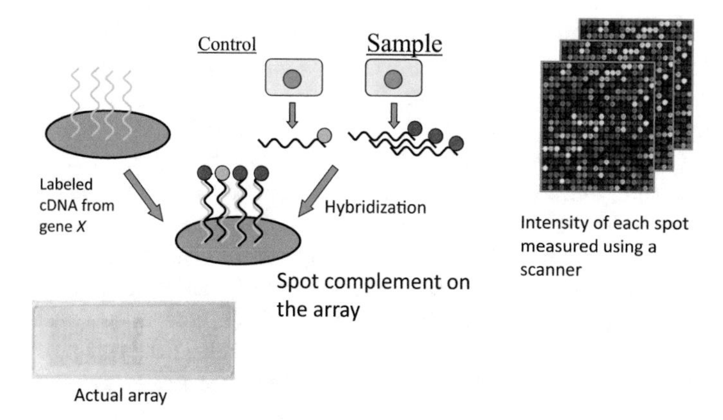

Fig. 5.2 Technologies for measuring gene expression

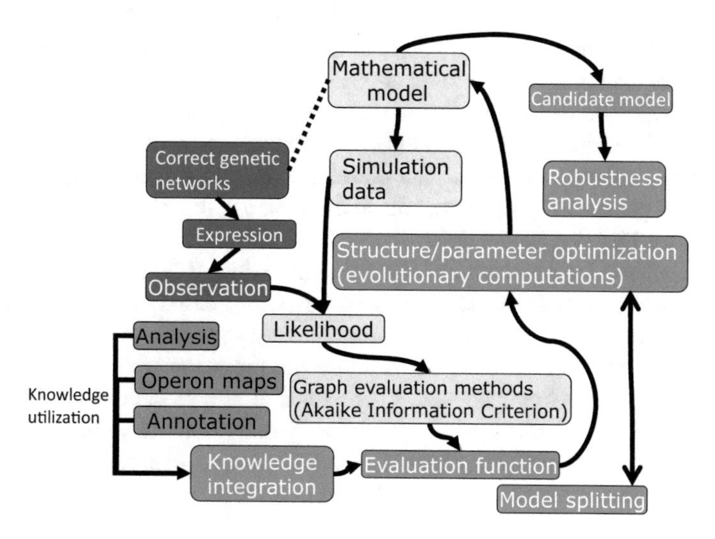

Fig. 5.3 Reverse engineering flowchart

This process of inferring and creating a structure from the given data is referred to as "reverse engineering" (Fig. 5.3). Design is a crucial process in most engineering fields, and design is also an interesting topic for artificial intelligence. This is because the process of design is a kind of inverse problem that involves finding a structure that matches a specification, and also because there is no general solution. In other words, sophisticated intelligence is necessary to generate a structure that satisfies the specifications while balancing competing criteria and carefully considering and modifying alternative candidates.

The ultimate objective of analyzing gene expression data is to elucidate the molecular mechanisms of genetic control networks. However, whether analysis succeeds depends on the accuracy and sensitivity of the experimental data for identifying the underlying biological system. If a dynamical model for genetic interactions can be found, then inferring a genetic network is essentially equivalent to learning the functional parameters in the structure.

Genetic networks are an important topic in bioinformatics. The goal of studying such systems is to decompose the systems that control the expression levels of each gene by modeling genes as reciprocally interacting networks. This is the inverse problem of inferring control routes and weights on the basis of the signals (substances) determined by a system of differential equations or a linear model. In this kind of design problem, the structure and parameters must be optimized simultaneously, and so tackling these problems using conventional methods, such as conjugate gradient descent or linear search, is infeasible. Because the molecular configuration is not well known, a limited amount of useful information is available, the constraints of the dynamic reactions must be satisfied, and the inverse problem is highly difficult.

There are different models that attempt to model GRNs [17, 18]. For instance, the popular S-System is proposed to model small networks. This model has been solved with different evolutionary computation techniques, and they have obtained good results. However, there are no models that achieve a perfect reconstruction of the network. We have implemented a variation of particle swarm optimization (PSO), called Dissipative PSO (DPSO), to optimize the model. We found that the combination of S-System and DPSO presents advantages against previously used methods and presents promising results to do inference in larger and more complex networks (see [19, 20] for details).

Several other AI approaches are also being investigated for the inference of gene regulatory networks (GRNs). Of these methods, those based on evolutionary computations, such as GA or GP, are becoming increasingly mainstream. By means of evolutionary computations, both the structure and the parameters can be optimized simultaneously, and there is no need to provide a prototype in advance. The search space for genetic networks contains numerous locally optimal solutions, so multipoint search methods are advantageous. The inference of gene regulatory networks using evolutionary computations is explained in great detail in [7].

Deep learning methods have also been applied to these inference problems. For example, one study [4] used deep learning to infer the expression of the gene of interest from the expression levels of landmark genes. When this method was verified against the microarray-based Gene Expression Omnibus dataset,[5] the results obtained were superior to those obtained by conventional methods such as linear regression. Another example is DeepBind [2], which uses convolutional neural networks to predict which DNA- and RNA-binding proteins control genes. There is also a study [15] that uses belief networks (see Sect. 2.3.3).

Genetic network inference has been used as a benchmark problem for AI and machine learning methods. Of these, a contest known as the DREAM project has been the focus of much attention. The main goal of DREAM (Dialogue for Reverse Engineering Assessments and Methods[6]) is to encourage integrated researches between

[5] http://dreamchallenges.org/challenges/.

[6] http://dreamchallenges.org/challenges/.

network inference, model construction theory, and wet experiments. DREAM is attempting to evaluate the quality of the experimental predictions and notations for the networks underlying biological systems. Researchers use a variety of optimization algorithms and AI to infer the structures of biological networks and to predict their robustness. The results are assessed according to numerous criteria, and the degree of success of each methodology is published as a ranking.

The DREAM project has evaluated more than thirty network inferences relating to microarray data for bacteria such as E.coli and S.aureus, as well as for enzymes and from computer simulations. It has also been shown that even better performance can be achieved by integrating these various approaches [11]. This is similar to the wisdom-of-the-crowds approach, in that it arrives at a final result by integrating the results derived from heterogeneous inference methods. The final inference can be obtained by skillfully combining the strengths of each method, while the shortcomings of any particular method are compensated for by another method. The methods used include the following:

- Regression: Selecting transcription elements using a linear regression specific to the target gene, in conjunction with bootstrapping and other methods,
- Mutual information: Inferring network edges from mutual information and relatedness.
- Correlation: Inferring network edges on the basis of Pearson correlation coefficients or Spearman rank-correlation coefficients,
- Bayesian method: Optimizing Bayesian networks by a variety of methods (such as simulated annealing or bootstrapping),
- Other meta-heuristics: Using Z scores, evolutionary computations, random forest search methods, and so on.

By integrating these methods using democratic principles, researchers have succeeded in creating robust methodologies that yield good results. This has allowed them to achieve highly reliable network inferences, reaching about 50% accuracy for 1,700 transcription control markers for E. coli and S. aureus.

5.2.2 INTERNe: IEC-Based GRN Inference with ERNe

This section describes an algorithm that integrates interactive evolutionary computation (IEC, see Sect. 1.6) and Genetic Algorithms (GAs) for the GRN inference. We call this method as INTERNe (INTeractive evolutionary computation with ERNe). We employ ERNe model for the GRN representation (see Sect. 5.4).

In INTERNe, the GA searches for the optimum solution through crossover of individuals. Crossover is carried out by choosing the base individual and the counterpart parent, and changing each element of the base individual with that of the counterpart parent. Mutation changes each element randomly. During searching of

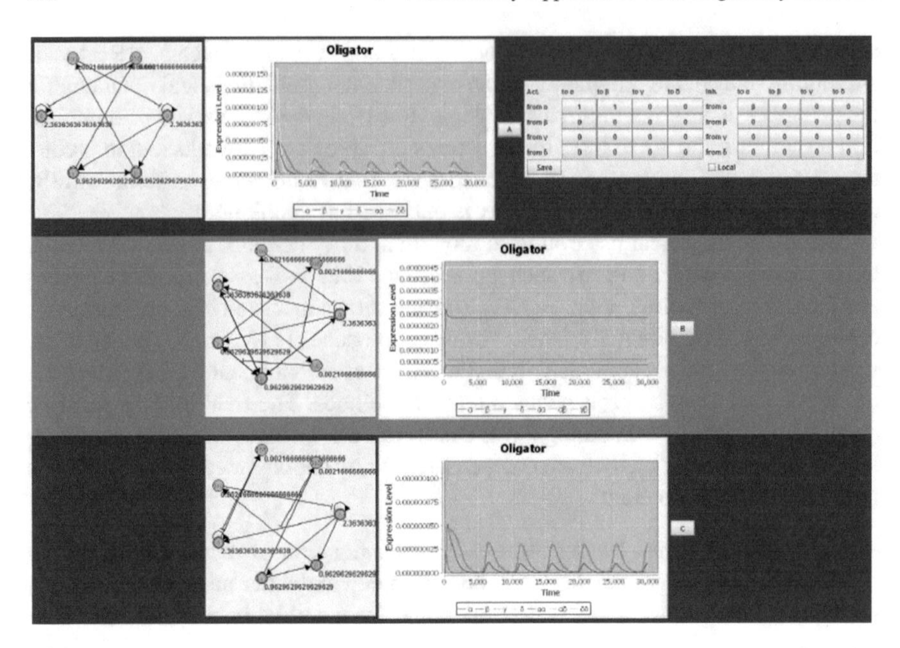

Fig. 5.4 IEC-based design for a GRN

genetic networks, the initial concentrations and reaction coefficients of the reaction model are searched using the standard values of the previous Ref.[13].

INTERNe optimizes genetic networks using both user evaluation and fitness functions in the following ways:

1. Creation of initial generation,
2. Evolutionary computation using fitness functions (GA phase),
3. Interactive evolutionary computation (IEC phase),
4. Return to GA phase.

We set the GA population to be 20. Crossover probability is set to be 0.3, whereas mutation probability is 0.1. The counterpart parent is chosen through the tournament method with size 2.

The IEC interface in INTERNe is shown in Fig. 5.4. The components that are 1 in the top right table are fixed during the search. Promotion from α to α itself and from α to β is fixed. The IEC user is expected to be a biologist with some expert knowledge, and these functions are expected to make the search smoother.

For the sake of the inference of sawtooth waves or Oligator (see 209 p.), the following two fitness functions are considered. First function takes into account the characteristics of the wave and is defined as follows:

$$fitness_1 = \frac{n^2}{max - min} \sum \left[(A_{max} - A_{min}) \frac{maxtomin}{mintomax} \right], \tag{5.1}$$

where max is the global maximum, min is the global minimum (of the wave curve), A_{max} are local maxima, A_{min} are local minima, $maxtomin$ is the time from local maximum to local minimum, $mintomax$ is the time from local minimum to local maximum, and n is the number of oscillations.

The second fitness function $fitness_2$ is defined as follows:

$$fitness_2 = \frac{limitcycle \times SDE}{MSE}, \tag{5.2}$$

where the *limitcycle* is defined in the following equation:

$$limitcycle = \frac{1}{(1+ \mid firstpeak - lastpeak \mid /firstpeak)^2}, \tag{5.3}$$

and is the penalty against attenuation. Here, *firstpeak* is the first amplitude (difference between local maximum and minimum) and *lastpeak* is the last amplitude in the time-series data. *SDE* is the standard deviation error, and *MSE* is the mean square error, i.e.,

$$MSE = \frac{1}{T} \int_0^T (\hat{x} - x)^2 dt, \tag{5.4}$$

where x is the actual value and $\hat{x}$ is the target value. The amplitude and period in this experiment are automatically determined after obtaining the waveform.

Figure 5.5 shows the waveforms evolved by INTERNe. For the sake of comparison, the same networks were designed using normal GAs. The population in this GA is 200, and the number of generations is about 50; individuals are evaluated about 10,000 times. INTERNe gave the desired waveform after several hundred evaluations in both cases; therefore, the number of evaluations is effectively reduced.

The following points should be considered to further improve INTERNe:

1. Further reduction of the number of evaluations is expected in order to avoid the user fatigue, e.g., less than several hundred evaluations and an evaluation time of one hour at most.
2. Relative evaluation (e.g., paired comparison) is desirable instead of absolute evaluation (e.g., where the user scores an individual).
3. The search space can be limited to leverage user insight, e.g., fixing some promotions and inhibitions by means of biological (expert) knowledge.

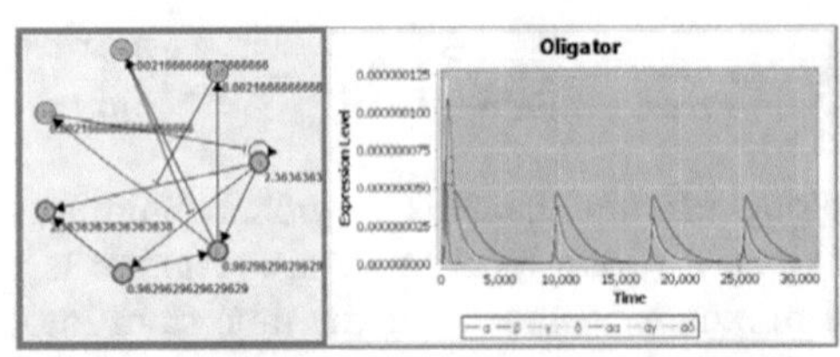

(a) Waveform generated by using $fitness_1$, GA: generation 31; IEC: generation 4.

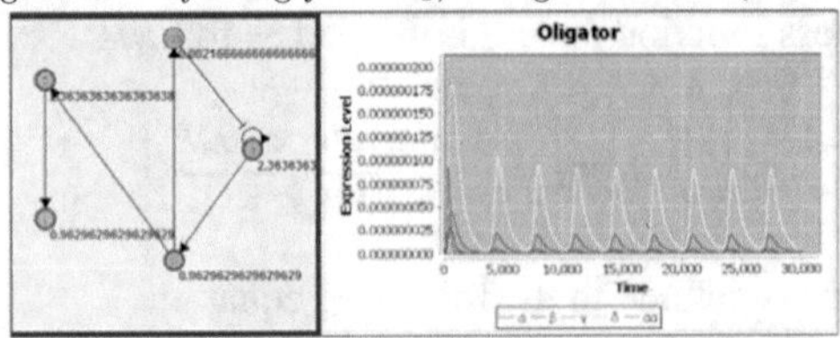

(b) Waveform generated by using $fitness_1$, GA: generation 39; IEC: generation 5.

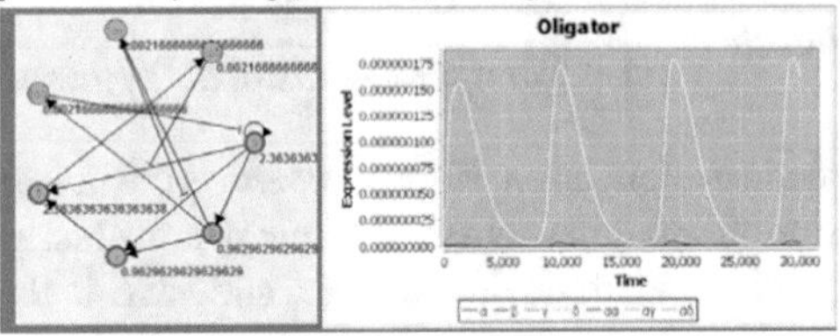

(c) Sawtooth wave generated by using $fitness_2$, GA: generation 39; IEC: generation 2.

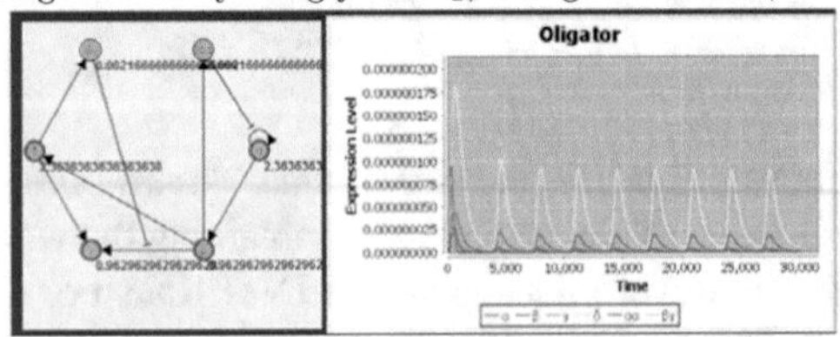

(d) Sawtooth wave generated by using $fitness_2$,GA: generation 47; IEC: generation 4.

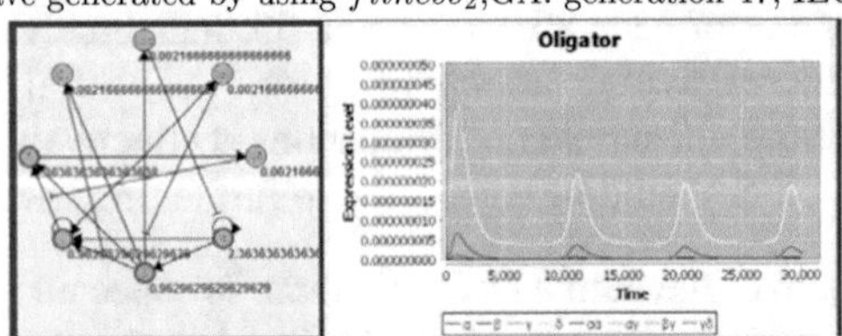

(e) Triangular wave generated by using $fitness_2$,GA: generation 25; IEC: generation 3.

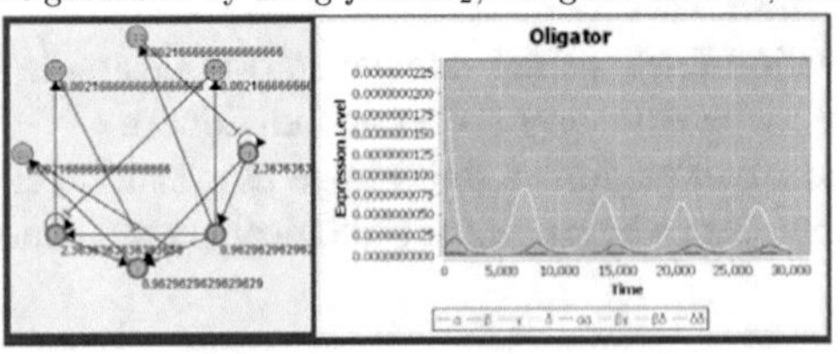

(f) Triangular wave generated by using $fitness_2$,GA: generation 33; IEC: generation 3.

Fig. 5.5 Evolved GRNs

5.3 MONGERN: GRN Application for Humanoid Robots

5.3.1 *Evolutionary Robotics and GRN*

Evolutionary robotics [14] is a method for producing autonomous robots via evolutionary processes. In other words, evolutionary robotics is an attempt to train appropriate robot behavior by representing the robot control system as a genotype and then applying EA search techniques.

Evolutionary robotics was proposed in 1993 by Cliff, Harvey, and Husbands [5], but evolutionary simulations of artificial objects with sensor motor systems were being conducted in the latter half of the 1980s. At the time, the control of production robots required special skills and advanced programming abilities. In a clear departure from traditional robotic design, a new generation of robots appeared in the 1990s that shared features of simple biological systems. These features are robustness, simplicity, compactness, and modularity. Those robots were designed to be easily programmed and controlled, even by people with backgrounds quite different from that of a conventional robot technician.

In recent years, evolutionary robotics has been the focus of attention from researchers in a variety of fields, including artificial intelligence, robotics, biology, and social behavioral science. In addition, evolutionary robotics shares many features with other AI approaches. These include

- Behavior-based robots,
- Robot learning,
- Multi-agent learning and cooperative learning,
- Artificial life,
- Evolvable hardware.

For example, Floreano et al. [14] have verified the Red Queen effect[7] through an evolutionary experiment with two Khepera robots (predator and prey). Each robot has six infrared sensors at the front and another two at the back. The predator is also equipped with visual sensors. The maximum speed of the prey is twice that of the predator. If the prey can run away cleverly, it will never be caught by the predator. Against this, the predator has more sensor information available to track the prey. The behavior of the predator and the prey was represented as genotypes via EANN direct coding (see 77 p.) and coevolved. One of the indexes used to characterize fitness was how well the prey survived. In other words, the longer the prey can evade capture by the predator, the greater the prey's fitness. Conversely, the faster the predator can catch the prey, the greater its fitness. The result of coevolution was the successive appearance of a series of fascinating tracking and evasion maneuvers, just as predicted by the Red Queen hypothesis. Examples of these behaviors include

[7] Red Queen effect is an evolutionary type of arms race. The name is derived from Alice's Adventures in Wonderland. See also 4 p.

the "spider strategy" (backing away slowly until a wall is visible and then catching the prey as it approaches) and circling one another like a dance.

There is growing demand for intelligent robots that can do various tasks in diverse environments. Humanoid robots have the same form as humans, and so can work in the same environments as humans without needing to change the environment, and can do dangerous work instead of humans in various situations. Such robots can therefore be used to perform a wider range of tasks than robots developed for a specific objective and are attracting much research interest.

When controlling the motion of humanoid robots, it is necessary to handle time-series data of the angles of joints. These robots have many degrees of freedom and must be able to move while the entire robot balances on two feet. Complex calculations of mechanics are necessary to obtain the angles of joints, and it is difficult to control motion. Therefore, research is being carried out to learn and generate the motion of robots using neural networks or reinforcement learning, enabling motion to be controlled without specifying all motion data.

There are high expectations for the application of gene regulatory networks (GRNs) to robotics because of their intrinsic robustness and modularity as well as the possibility of evolution and compression of complex expression. Research using GRN is already being conducted in the field of robotics to form patterns by various robots, and robustness has been demonstrated [9].

The drawback of humanoid robots is that walking on two feet is unstable and is easily disturbed, and thus the GRNs require robustness for a network to generate the motion of humanoid robots. Generating the motion of humanoid robots using robust GRNs may generate humanoid robot motion that is more robust against disturbances than conventional methods. Thus, applying GRN to a complex task in a real problem, i.e., control of humanoid robots, would demonstrate that GRNs can flexibly learn and are robust.

The sections describe learning of GRNs through differential evolution for learning of networks that generate multiple motion and verify whether the networks generate motion correctly using simulators and actual robots. We call this system as MON-GERN (MOtioN generation by GEne Regulatory Networks).

5.3.2 How to Express Motions

The motion of a robot can be expressed by time-series motion of joints. Therefore, GRNs can express the motion of a robot by relating the output values from GRNs to the angles of each joint. Humanoids have many degrees of freedom, the generation and memory of motion are difficult, and hence methods of using models to abstract motion are widely researched. Models employed include hidden Markov models [8] and neural networks [22], and the values output from nodes is typically related to angles of joints.

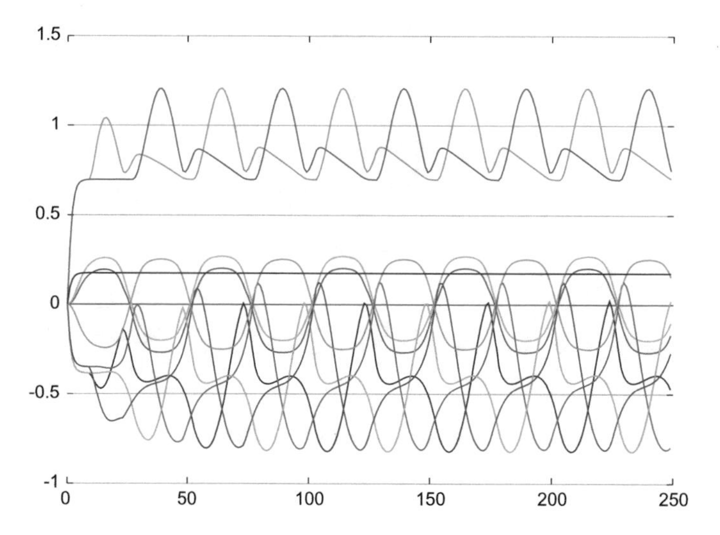

Fig. 5.6 Time-series data of walking motion of a humanoid robot

When using GRNs for humanoid robot motion generation and learning models, it is necessary to convert the output from the networks to data that express the motion of robots. The motion of humanoid robots can usually be expressed as time-series data of angles of joints. Figure 5.6 shows an example of the walking motion of humanoid robots. Each time series corresponds to the angle of each joint. Therefore, motion can be generated by generating such time-series data. As mentioned, each gene outputs time-series data in GRNs; therefore, by feeding this output into robots, GRNs can generate motion. GRNs are used to generate the motion of humanoid robots by relating the output from each gene to each joint of a humanoid robot.

Figure 5.7 shows the method of generating motion in MONGERN. The time-series data output from the network is related to joints and converted to joint angles, and the robot is controlled by providing joint angles as input.

In MONGERN, the output of a GRN is converted to joint angles as follows. Each gene of the GRN must first be related to each joint of the humanoid robot. The number of moving joints in the humanoid robot is limited, and the genes of the GRN are related to the moving joints. In principle, one gene is related to one joint, but learning costs are reduced by relating one gene to two joints that perform symmetric motion on each side. The angles of joints to which no gene in the network is related are always fixed to a certain value during motion.

The values output from the network is converted based on the designated method for each related joint. The outputs from GRNs, which are the concentration of proteins, are always positive, and thus there are limitations on the range that joints can move if directly used as joint angles. Therefore, MONGERN employs a linear transformation that allows motion over the entire range of movement for each joint of the robot, and therefore there are no limits on the motion.

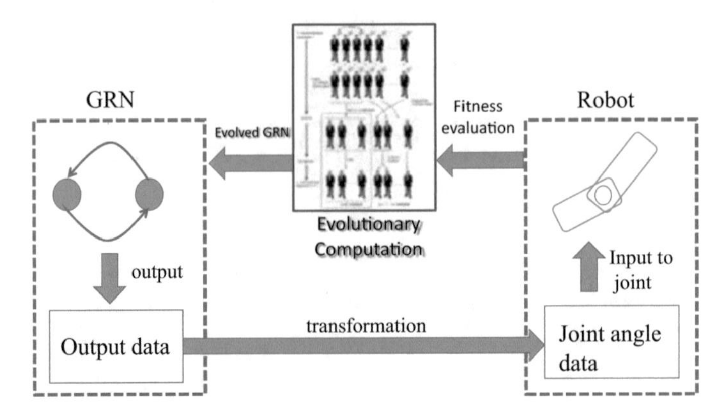

Fig. 5.7 Motion generation system for humanoid: MONGERN

5.3.3 How to Learn Motions

The artificial gene network (AGN) is utilized as a model to express GRNs.

In an AGN [12], the interrelationships of genes in gene networks are represented by the following simultaneous differential equations:

$$\frac{dG_i}{dt} = a_i \cdot \prod_j \left(\frac{K_{I_j}^n}{I_j^n + K_{I_j}^n} \right) \cdot \prod_k \left(\frac{A_k^n}{A_k^n + K_{A_k}^n} \right) - b_i \cdot G_i, \tag{5.5}$$

where i represents the number of the gene and the AGN becomes a simultaneous differential equation that grows with the total number of genes. The value of G_i is the output of the gene control network and indicates protein concentration. Therefore, the left side represents the concentration change of the protein synthesized by the gene. For the right-hand side, I (representing an inhibitor) expresses the concentration of the protein that suppresses the expression of the ith gene, and A (representing the activator) expresses the concentration of the protein that promotes the expression of the ith gene. The first term on the right side expresses the synthesis of the protein, and the second expresses protein decomposition. The parameters a_i and b_i express the synthesis rate and decomposition rate, respectively. The constants K_{Ij} and K_{Ak} represent concentrations at which the effect of the inhibitor or activator is half of its saturating value. The exponent n regulates the sigmoidicity of the curve.

The network structure can be determined using AGN parameters, so that the parameters shown in Table 5.1 are used as genotypes in MONGERN. In other words, GRNs are searched by adjusting parameters $K_{Ij}, K_{Ak}, n, a_i, b_i$. Table 5.2 shows the ranges of parameters used. When parameters fall outside this range after genetic operations such as crossover, parameters exceeding the maximum value are set to twice the maximum value minus the original value, and those less than the minimum value are set to twice the minimum value minus the original value.

Table 5.1 Genotype (parameters in AGN)

n	a_1 ...	b_1 ...	K_{I1} K_{I2} ... K_{A1} K_{A2} ...
	$\Leftarrow$ #.of genes $\Rightarrow$	$\Leftarrow$ #.of genes $\Rightarrow$	$\Leftarrow$ (#.of genes) $\times$ 2 $\Rightarrow$

Table 5.2 Range of AGN parameters

Parameter	Maximum value	Minimum value
n	1.5	5.0
a	0.0	2.0
b	0.02	0.15
K	−1.0	1.0

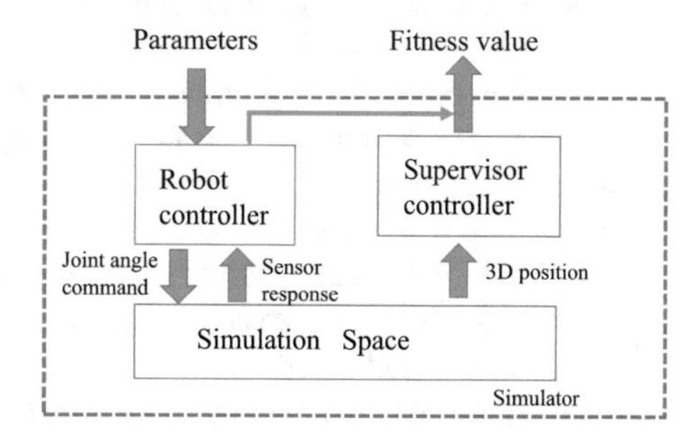

Fig. 5.8 Evaluation of individuals in GRN-based motion learning

MONGERN uses differential evolution (DE, see Sect. 2.1.2) to learn the robot motion. To apply differential evolution to GRNs, the structure of the network must be converted to numerical values and then expressed as genotypes that can be used for DE. First, parameter sequences of the GRN (Table 5.1) are generated as initial individuals and each individual is evaluated. The parameters are randomly generated within the ranges shown in Table 5.2. Genetic operations are carried out through DE, and learning is performed by continuing to generate generations through fitness evaluation.

Individuals are evaluated using a simulator as shown in Fig. 5.8. The controller of the robot in the simulator calculates time-series data from parameter sequences of the network and controls the angles of the joints of the robot. The robot controller can obtain the angles of joints and the sensor values of the robot, and records values necessary for evaluation. A supervisor controller is run at the same time as the robot controller. The supervisor controller can obtain the three-dimensional positions of objects in the simulator space and records values for evaluation such as the positions of joints. The fitness is calculated once a simulation finishes based on values recorded by the controllers during the simulation. The fitness function is designated for each motion to evaluate the attainment of the motion.

5.3.4 How to Make Robust Motions

GRNs are known to be robust against changes in the environment. Here, verification of the robustness of GRNs means that the network correctly outputs the intended output even when parameters are changed or the system is disturbed. The motion generation method described in Sect. 5.3.2 does not incorporate sensor values that detect the outside environment; therefore, GRNs always output the same value independent of the external environment. This does not mean that the generated motion is leveraging the robustness of GRNs.

This section describes a method for generating robust motion that incorporates the response of joint angles as sensors. The output of GRNs is given as a relation of simultaneous differential equations. As a result, the output of the next step can be calculated by giving initial values and a time step, and time-series values can be generated by iterating such calculations. MONGERN receives joint angles of the robot at each step as the response, and the joint angles of the next step are generated not by using the previous output but by using the response of joint angles. The command to joints and the corresponding response may be different, such as when the robot hits an obstacle or a joint exceeds its range of movement. It is expected that environmental robustness can be achieved by incorporating the response as sensors.

In MONGERN, input values are first supplied to GRNs, and then the output for the next step is obtained, converted, and given to the robot as commands of joint angles. The responses of joint angles are obtained, converted, and input into networks, and then the output for the next step is obtained. Motion is generated by repeating this procedure.

In this phase, the learning method uses DE. Individuals are evaluated by carrying out simulations of an individual in many environments, and the total of the evaluated values is used as the evaluation value of this individual, i.e., the GRN. The most important factor that defines the robustness of a humanoid robot is the ability to move without falling over in different environments. The stability of humanoid robots may break down with a small shift of one joint; therefore, the robustness of the GRN is likely to not directly contribute to the robustness of humanoid robot motion. In this case, learning in one environment would not lead to robust motion. Learning in many environments will result in searching for robust networks.

In general, it is not good to continue to provide commands when the command and response of joint angles differ; control according to the situation is necessary. However, processing of such exceptions is difficult. Therefore, robust networks must be able to generate correct motion by autonomously changing when the response is difficult, and MONGERN should achieve this type of robustness.

Table 5.3 Specifications of HOAP-2

Height	About 50 cm
Weight	About 7 kg
Joint degrees of freedom	6 DOF/Leg × 2
	4 DOF/Arm × 2
	1 DOF/Waist × 1
	1 DOF/Hand × 2
	2 DOF/Neck × 1
Sensor	Joint angle sensor
	Three-axis acceleration sensor
	Three-axis angular velocity sensor
	Foot bottom pressure sensor

5.3.5 *Simulation Experiments with MONGERN*

The following experiments are carried out to demonstrate the effectiveness of MONGERN. We have used the humanoid robot HOAP-2 from Fujitsu Automation (see Figs. 5.12 and 5.13). The main specifications of HOAP-2 are shown in Table 5.3.

The motions learned are: (a) handstand, (b) walk, (c) bend and stretch, (d) rise, (e) kick, and (f) turn over. In all motion learning experiments, there are 20 individuals per generation, the scaling parameter for mutation in DE is 0.6, and the crossover rate is 0.9. The fitness functions and joints used in each motion are described below. Table 5.4 shows genes assigned to joints for each motion.

(a) Handstand

The fitness is the average of both feet of the distance between the toe and the ground at the end of the simulations. The joints used are the waist (front/back direction), hip (front/back direction for both right and left), knee (right and left), and ankle (front/back direction for both right and left). The same gene is used to express both right and left joints if applicable.

(b) Walk

The fitness is the average of both feet of the distance between the initial and final positions of the toe. When the robot becomes unstable during a simulation, the fitness is multiplied by a correction term of $\frac{\text{time until instability}}{\text{simulation time}}$. The joints used are the hip (front/back direction for both right and left), knee (right and left), and ankle (front/back direction for both right and left). Different genes are used for different joints because the right and left joints do not move symmetrically. To stabilize, the right and left hip joint angles (front/back direction) are given as (knee joint angle) − −(ankle joint angle).

Table 5.4 Assignment of genes for joints

Gene	(a) Handstand	(b) Walk
G0	Waist (front/back direction)	Left knee
G1	Hip (front/back direction for both right and left)	Right knee
G2	Knee (both right and left)	Left ankle (front/back direction)
G3	Ankle (front/back direction for both right and left)	Right ankle (front/back direction)
Gene	(c) Bend and stretch	(d) Rise
G0	Hip (front/back direction for both right and left)	Waist (front/back direction)
G1	Knee (both right and left)	Hip (front/back direction for both right and left)
G2	Ankle (front/back direction for both right and left)	Knee (both right and left)
G3	None	Ankle (front/back direction for both right and left)
G4	None	Shoulder (front/back direction for both right and left)
G5	None	Shoulder (right/left direction for both right and left)
Gene	(e) Kick	(f) Turn over
G0	Right hip (right/left direction)	Left hip (torsion)
G1	Right ankle (right/left direction)	Left hip (right/left direction)
G2	Right hip (front/back direction)	Left hip (front/back direction)
G3	Right knee	Left knee
G4	Right ankle (front/back direction)	Right shoulder (front/back direction)

(c) Bend and Stretch

Key frames where the robot is sitting down and standing up are prepared, and the joint angles are considered as standard data. The fitness is taken as the sum of square errors in joint angles from these standard data. When the robot becomes unstable during a simulation, the fitness is multiplied by a correction term of $\frac{\text{time until instability}}{\text{simulation time}}$. The joints used are the hip (front/back direction for both right and left), knee (right and left), and ankle (front/back direction for both right and left). The same gene is used to express both right and left joints if applicable.

(d) Rise

The fitness is the average of foot sensor values of both feet. The joints used are the waist (front and back), hip (front/back direction for both right and left), knee (right and left), ankle (front/back direction for both right and left), and shoulder (both front/back direction and right/left direction for both right and left). The same gene is used to express both right and left joints if applicable.

(e) Kick

The fitness is taken as the maximum speed of the ball during the simulation. When the robot becomes unstable during a simulation, the fitness is multiplied by a correction term of $\frac{\text{time until instability}}{\text{simulation time}}$. The ball in Environment 1 (diameter 14 (cm), distance 12 (cm), etc.) in Table 5.5 is used. The joints used are the right hip (right/left direction), right ankle (right/left direction), right hip (front/back direction), right knee, and right ankle (front/back direction). Different genes are used for different joints.

(f) Turn Over

Taking the x-axis as the line connecting both shoulder joints, the robot is laid down face up such that the x coordinate of the left shoulder joint is larger than that of the right shoulder. The fitness is given as follows:

$$x \text{ coord. of the right shoulder joint} - x \text{ coord. of the left shoulder joint.}$$

The joints used are the left hip (torsion), left hip (right/left direction), left hip (front/back direction), left knee, and right shoulder (front/back direction). Different genes are used for different joints.

The results of learning of each motion are shown below.

(a) Handstand Motion

The results of learning the handstand motion are shown in Fig. 5.9a. The robot does a handstand after both arms and the head touch the ground. One thousand and five hundred generations are necessary to learn the motion. Figure 5.10a shows time-series data output by the gene regulatory network, and the structure of the GRN is given in Fig. 5.11a.

The evolved AGN simultaneous differential equations are as follows.

$$\frac{dG_0}{dt} = 1.8 \cdot \left(\frac{1^2}{G_0^2 + 1^2}\right) \cdot \left(\frac{G_1^2}{G_1^2 + 0.5^2}\right) \cdot \left(\frac{0.9^2}{G_2^2 + 0.9^2}\right) \cdot \left(\frac{0.4^2}{G_3^2 + 0.4^2}\right) - 0.1G_0$$

$$\frac{dG_1}{dt} = 2 \cdot \left(\frac{1^2}{G_0^2 + 1^2}\right) \cdot \left(\frac{1^2}{G_1^2 + 1^2}\right) \cdot \left(\frac{1^2}{G_3^2 + 1^2}\right) \cdot \left(\frac{0.6^2}{G_4^2 + 0.6^2}\right) - 0.1G_1$$

$$\frac{dG_2}{dt} = 0.4 \cdot \left(\frac{0.9^2}{G_0^2 + 0.9^2}\right) \cdot \left(\frac{0.3^2}{G_1^2 + 0.3^2}\right) \cdot \left(\frac{0.5^2}{G_2^2 + 0.5^2}\right) \cdot \left(\frac{G_3^2}{G_3^2 + 0.7^2}\right) \cdot \left(\frac{0.5^2}{G_4^2 + 0.5^2}\right)$$

$$\frac{dG_3}{dt} = 1.9 \cdot \left(\frac{G_0^2}{G_0^2 + 0.3^2}\right) \cdot \left(\frac{0.4^2}{G_2^2 + 0.4^2}\right) \cdot \left(\frac{1^2}{G_3^2 + 1^2}\right) \cdot \left(\frac{0.7^2}{G_4^2 + 0.7^2}\right) - 0.1G_3$$

$$\frac{dG_4}{dt} = 1.8 \cdot \left(\frac{0.9^2}{G_0^2 + 0.9^2}\right) \cdot \left(\frac{1^2}{G_1^2 + 1^2}\right) \cdot \left(\frac{G_2^2}{G_2^2 + 0.6^2}\right) \cdot \left(\frac{1^2}{G_3^2 + 1^2}\right) \cdot \left(\frac{0.3^2}{G_4^2 + 0.3^2}\right) - 0.1G_4$$

(b) Walk Motion

The results of learning the walk motion are shown in Fig. 5.9b. The robot is dragging its legs, but positions each foot forward in turn. Three thousand generations are

Fig. 5.9 Simulation results

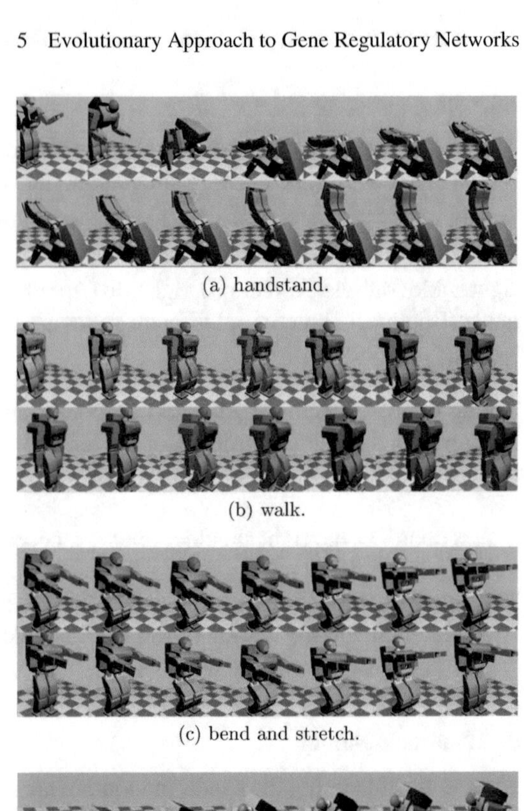

(a) handstand.

(b) walk.

(c) bend and stretch.

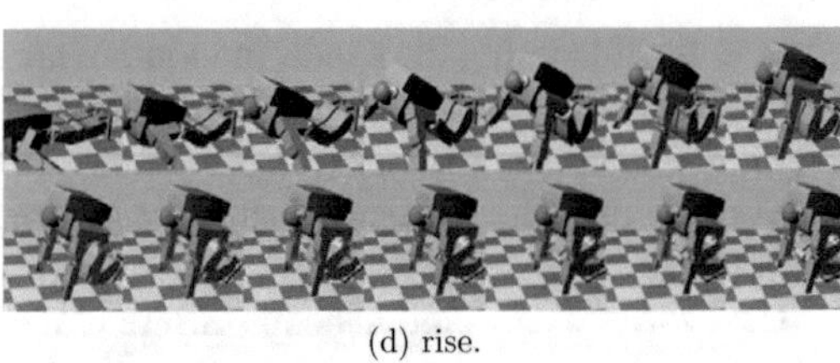

(d) rise.

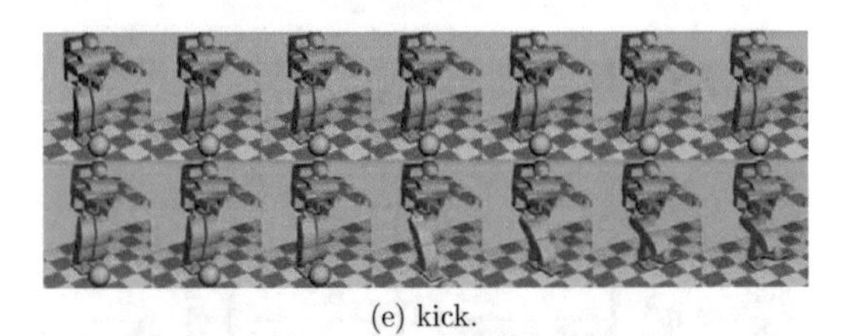

(e) kick.

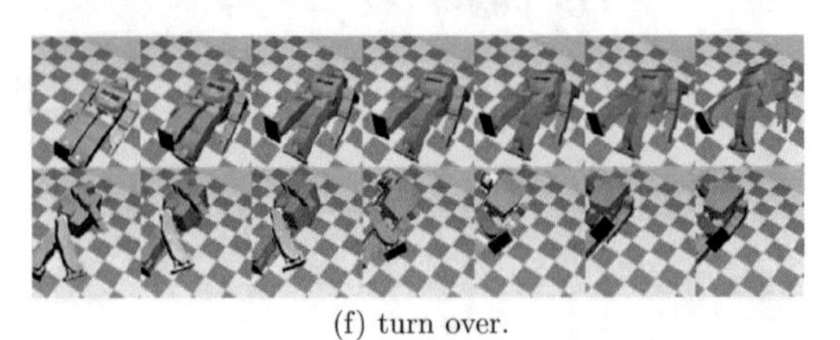

(f) turn over.

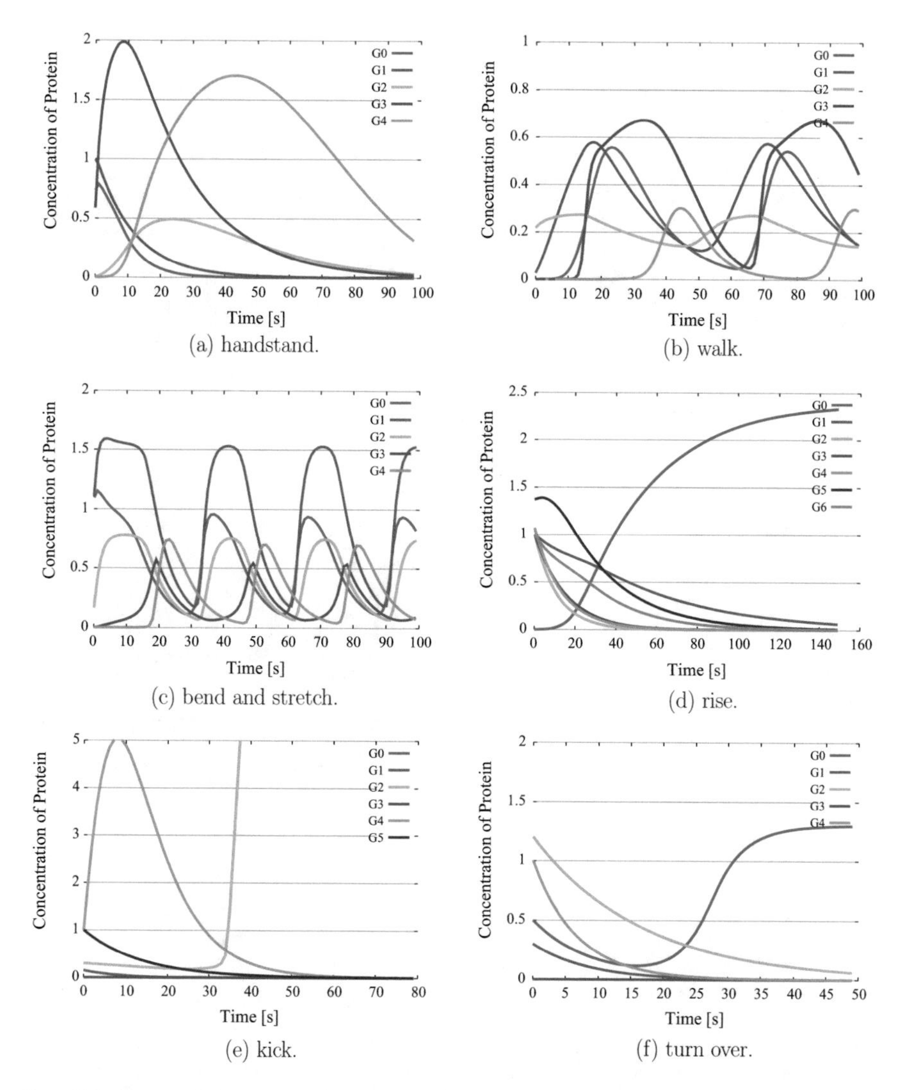

(a) handstand.
(b) walk.
(c) bend and stretch.
(d) rise.
(e) kick.
(f) turn over.

Fig. 5.10 Output of the gene regulatory network (GRN)

necessary to learn the motion. Figure 5.10b shows time-series data output by the gene regulatory network, and the structure of the GRN is given in Fig. 5.11b.

(c) Bend and Stretch Motion

The results of learning the bend and stretch motion are shown in Fig. 5.9c. The robot bends and stretches stably. Two thousand and three hundred generations are necessary to learn the motion. Figure 5.10c shows time-series data output by the gene regulatory network, and the structure of the GRN is given in Fig. 5.11c.

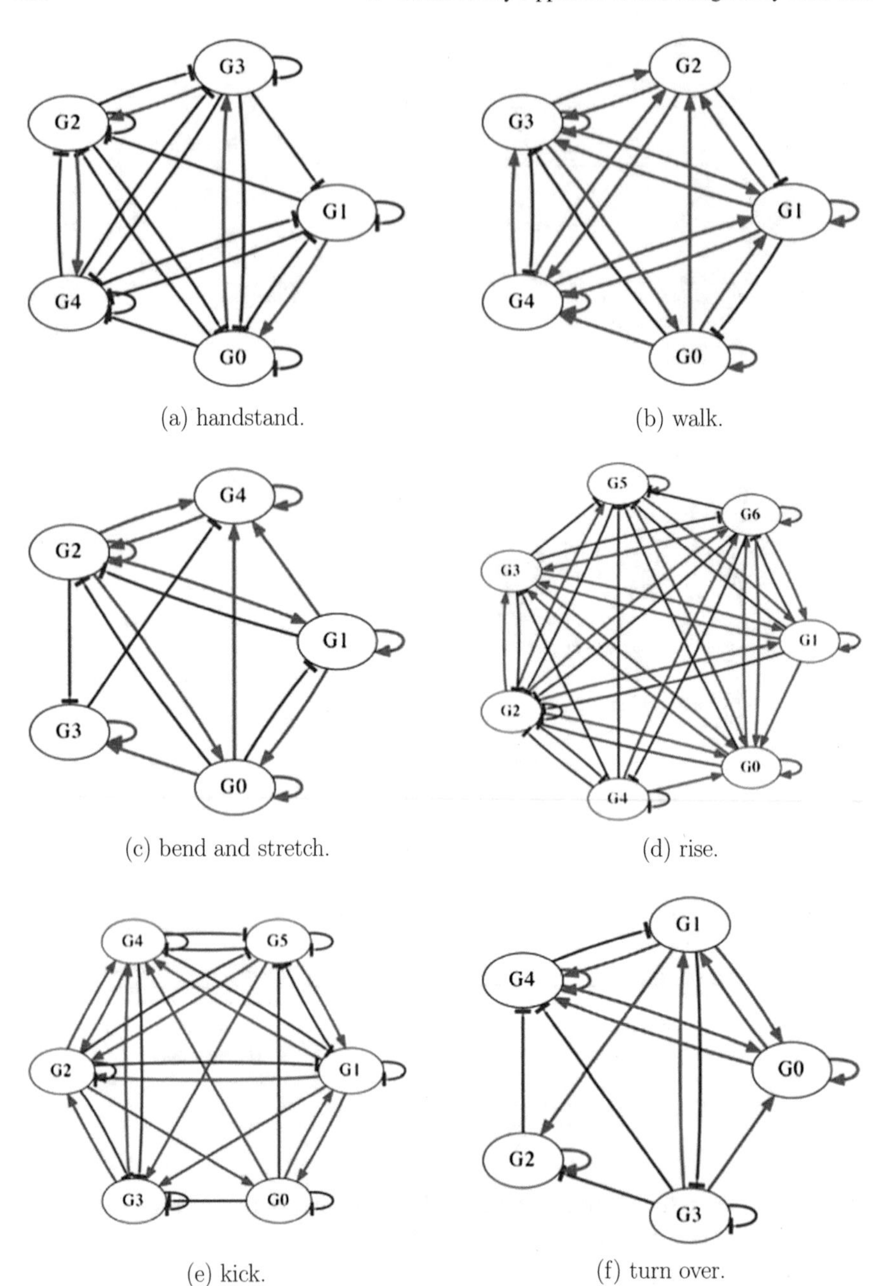

(a) handstand.

(b) walk.

(c) bend and stretch.

(d) rise.

(e) kick.

(f) turn over.

Fig. 5.11 Structures of the gene regulatory network (GRN)

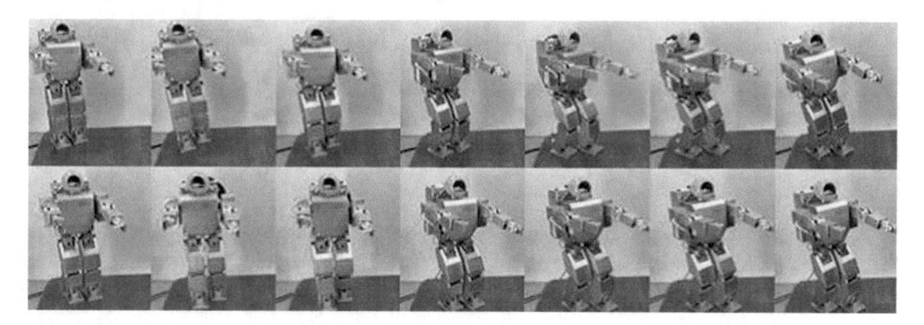

Fig. 5.12 Verification using an actual robot of (b) walk

(d) Rise Motion

The results of learning the rise motion are shown in Fig. 5.9d. The robot bends and stretches stably. Nine hundred and sixty generations are necessary to learn the motion. Figure 5.10d shows time-series data output by the gene regulatory network, and the structure of the GRN is given in Fig. 5.11d.

(e) Kicking Motion

The results of learning the kicking motion are shown in Fig. 5.9e. The robot is kicking a ball. One thousand and two hundred generations are necessary to learn the motion. Figure 5.10e shows time-series data output by the gene regulatory network, and the structure of the GRN is given in Fig. 5.11e.

(f) Turn Over Motion

The results of learning the turn over motion are shown in Fig. 5.9f. The robot turns over from face up to face down. Six hundred generations are necessary to learn the motion. Figure 5.10f shows time-series data output by the gene regulatory network, and the structure of the GRN is given in Fig. 5.11f.

5.3.6 Real Robot Experiments with MONGERN

The networks obtained using motion learning experiments are used to operate a real HOAP-2 robot. The results of (b) walk and (c) bend and stretch motions are shown in Figs. 5.12 and 5.13, respectively. The robot moves without falling in both cases, showing that it correctly moves in a real environment. In walk motion, the entire robot faces along the leg that moves forward, and the distance of each step is shorter than in the simulation (see Fig. 5.9b).

5.3.7 Robustness with MONGERN

Learning of the motion of kicking a ball is carried out to verify the robustness of GRNs evolved by MONGERN. Kicking motion requires robustness because the

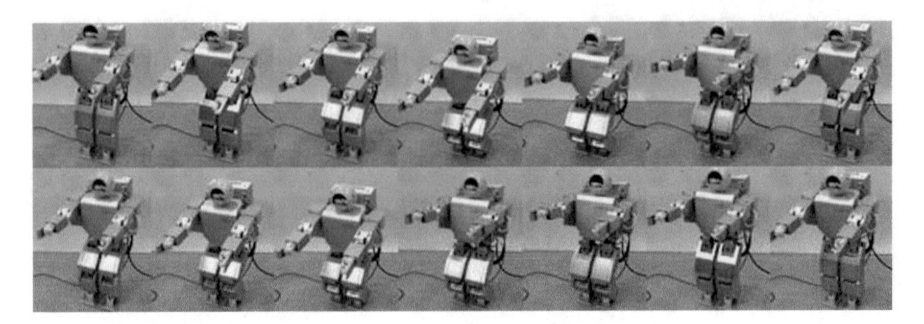

Fig. 5.13 Verification using an actual robot of (c) bend and stretch

Table 5.5 Experimental environments to learn robust kicking motion

Environment id.	Type	Density (kg/m^3)	Diameter (cm)	Distance (cm)
1	Ball	1000	14	12
2	Ball	3000	30	26
3	Ball	10,000	18	11
4	Ball	10,000	24	12
5	Ball	2000	40	20
6	Block	2000	–	–
7	None	–	–	–

motion involves standing on one leg and the force acting on the robot changes when the ball is changed. This experiment performs evaluations for the same individual under multiple environments where different objects are kicked. The objects kicked in each environment are summarized in Table 5.5. To adapt to many conditions using a small number of environments, two extreme environments are employed: Environment 6 that consists of a block corresponding to a wall, and Environment 7 where no object is placed and the robot kicks nothing. The dimensions of the block are height 15 cm, width 30 cm, and depth 30 cm. The distances given in this table are the distance from the center of the object to the toe of the robot. The fitness in each environment is the highest speed of the ball $\times$ $\frac{\text{time until instability}}{\text{simulation time}}$. The fitness of seven environments is added to obtain the total fitness of the networks for the individual. The joints used are the left ankle (right/left direction), right hip (front/back direction), right ankle (front/back direction), and left hip (front/back direction). Different genes are used for different joints. Genes are assigned to joints as shown in Table 5.6.

The results of simulations in Environment 1 are shown in Fig. 5.14. Figure 5.15 shows time-series data output by the GRN after learning, and the structure of the GRN is given in Fig. 5.16. The time-series data for Environment 6 with the same network is shown in Fig. 5.17. Kicking the heavy block corresponding to a wall bends the left hip joint (G3), which inhibits the right hip joint (G1).

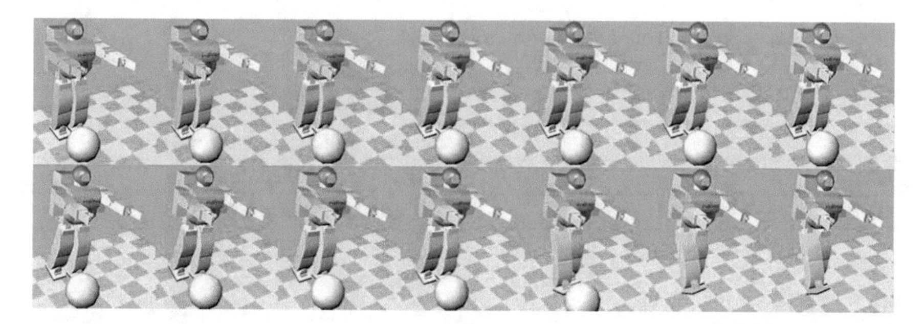

Fig. 5.14 Simulation results of kicking motion in Environment 1

Fig. 5.15 Output of the gene regulatory network (GRN) after learning robust kicking motion in Environment 1

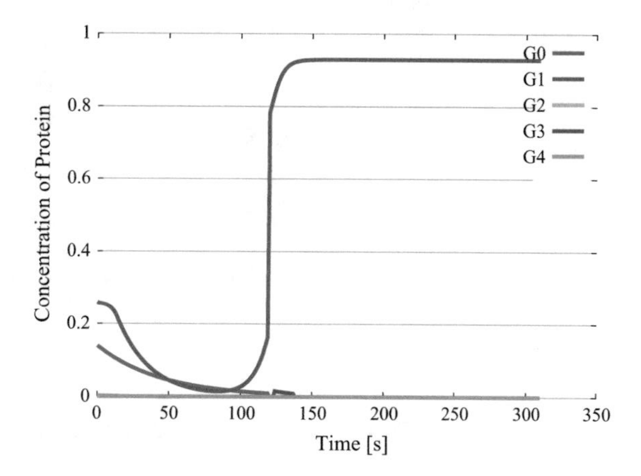

Fig. 5.16 Structure of the gene regulatory network (GRN) after learning robust kicking motion

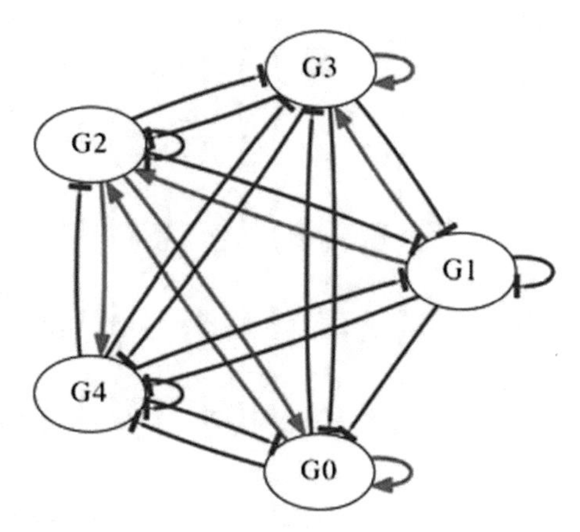

Table 5.6 Assignment of genes for joints in robust kicking motion

Gene	Assigned joint
G0	Left ankle (right/left direction)
G1	Right hip (front/back direction)
G2	Right ankle (right/left direction)
G3	Left hip (front/back direction)

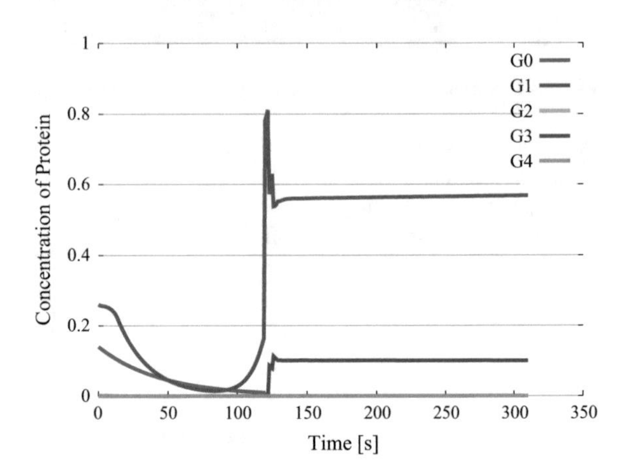

Fig. 5.17 Output of the gene regulatory network (GRN) after learning robust kicking motion in Environment 6

The evolved AGN simultaneous differential equations are as follows.

$$\frac{dG_0}{dt} = 1.9 \cdot \left(\frac{G_0^5}{G_0^5 + 0.9^5}\right) \cdot \left(\frac{0.7^5}{G_1^5 + 0.7^5}\right) \cdot \left(\frac{G_2^5}{G_2^5 + 0.8^5}\right) \cdot \left(\frac{0.1^5}{G_3^5 + 0.1^5}\right) \cdot \left(\frac{0.4^5}{G_4^5 + 0.4^5}\right) - 0.1G_0$$

$$\frac{dG_1}{dt} = 1.7 \cdot \left(\frac{0.5^5}{G_1^5 + 0.5^5}\right) \cdot \left(\frac{0.6^5}{G_2^5 + 0.6^5}\right) \cdot \left(\frac{0.1^5}{G_3^5 + 0.1^5}\right) \cdot \left(\frac{0.7^5}{G_4^5 + 0.7^5}\right) - 0.1G_1$$

$$\frac{dG_2}{dt} = 0.6 \cdot \left(\frac{G_0^5}{G_0^5 + 0.2^5}\right) \cdot \left(\frac{G_1^5}{G_1^5 + 0.6^5}\right) \cdot \left(\frac{0.6^5}{G_2^5 + 0.6^5}\right) \cdot \left(\frac{0.9^5}{G_3^5 + 0.9^5}\right) \cdot \left(\frac{0.3^5}{G_4^5 + 0.3^5}\right) - 0.1G_2$$

$$\frac{dG_3}{dt} = 1.8 \cdot \left(\frac{0.2^5}{G_0^5 + 0.2^5}\right) \cdot \left(\frac{G_1^5}{G_1^5 + 0.3^5}\right) \cdot \left(\frac{0.9^5}{G_2^5 + 0.9^5}\right) \cdot \left(\frac{G_3^5}{G_3^5 + 0.4^5}\right) \cdot \left(\frac{0.5^5}{G_4^5 + 0.5^5}\right) - 0.1G_3$$

$$\frac{dG_4}{dt} = 1.7 \cdot \left(\frac{0.8^5}{G_0^5 + 0.8^5}\right) \cdot \left(\frac{0.9^5}{G_1^5 + 0.9^5}\right) \cdot \left(\frac{G_2^5}{G_2^5 + 0.5^5}\right) \cdot \left(\frac{0.8^5}{G_3^5 + 0.8^5}\right) \cdot \left(\frac{0.7^5}{G_4^5 + 0.7^5}\right) - 0.1G_4$$

Robustness of the evolved GRN is verified by investigating stability with various balls other than those used in learning. A combination of 42 environments is used in the motion verification experiments, where six ball densities of 1000, 2000, 3000, 5000, 7000, and 10,000 (kg/m^3) and seven ball diameters of 10, 15, 20, 25, 30, 35, and 40 (cm) are used. Here, the ball radius is the distance between the center of the ball and the tip of the robot. We have confirmed that the robot successfully kicked the ball stably in all 42 combinations.

Fig. 5.18 Trap motion evolved by GRN

We have conducted comparative experiments with other traditional methods. We chose two target motions for the sake of comparison. These are headstands and trap motions (see Fig. 5.18). In this experiment, we try to evolve the chest trap motion, which is one of four primary ways to trap a soccer ball (the other ways are the head trap, the thigh trap, and the foot trap). MONGERN successfully generated trap motions as shown in Fig. 5.18. On the other hand, it has been known that this trap motion is very hard to be generated by means of traditional methods, e.g., neural networks and reinforcement learning, which have been frequently used for robot motion learning. This is partly because the trapping refers to the process of stopping or slowing a ball that has traveled toward a robot, which is accompanied with discrete and abrupt changes. Thus, we have confirmed the effectiveness of MONGERN in this context.

5.4 ERNe: A Framework for Evolving Reaction Networks

In this section, we explain ERNe (Evolving Reaction Network[8]), a mechanism that expands the NEAT methodology (see Sect. 3.1.1) to produce a type of genetic network suitable for biochemical systems [6]. ERNe is a methodology that focuses on the similarity between neural networks and biochemical networks, using evolutionary methods to train the network structure in silico. ERNe has succeeded in designing genetic circuits in synthetic biology. In fact, ERNe is faster than conventional methods at discovering known oscillatory circuits [13], bistable networks [16], and oscillatory circuits exhibiting complex behavior. The efficacy of ERNe has also been demonstrated through application to a variety of search problems and to living systems.

5.4.1 DNA PEN Toolbox

DNA PEN (Dynamic Network Assembly Polymerase Exonuclease Nickase) Toolbox is inspired by gene regulatory networks [13]. In such networks, regulation comes

[8]Erne is also another name for a sea eagle.

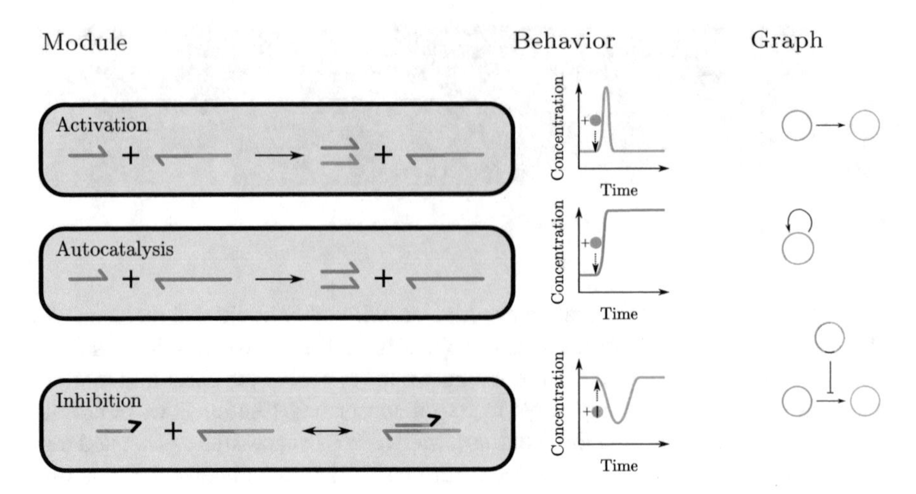

Fig. 5.19 The modules of the DNA PEN Toolbox: activation, autocatalysis and inhibition

in two flavors: activatory or inhibitory, respectively, increasing and decreasing the expression of a targeted gene. One of the goals is to reproduce such mechanisms while removing the complexity of using RNA, proteins, and the associated molecular machinery.

The toolbox mimics regulatory mechanisms through specific DNA species, activating or inhibiting the generation of other DNA species. Activators are often referred to as signal, while DNA species performing inhibition are simply called inhibitors. Activation is done through the help of a second class of DNA species called templates. At a coarse-grained level, templates are DNA species present in the solution that generates an output from an input, eventually releasing both. The output can itself be an activator, allowing the cascading of multiple reactions. In the DNA PEN Toolbox, the exact mechanism through which activation is done is based on a combination of polymerase and nickase enzymes [13]. However, other mechanisms, such as Qian and Winfree's seesaw gate [21], could implement similar functions.

A schematic representation of both activation and inhibition, as well as the special case of autocatalytic activation, as implemented in the DNA PEN Toolbox is shown in Fig. 5.19.

At the lowest level of abstraction, template species are considered as black boxes (Fig. 5.20). Those boxes have a simple phenomenological transfer function inspired by Michaelis–Menten reaction rates and competitive inhibition. The rational behind this model is to assume that hybridization and denaturation (two complementary DNA strands attaching and detaching, respectively) reach equilibrium much faster

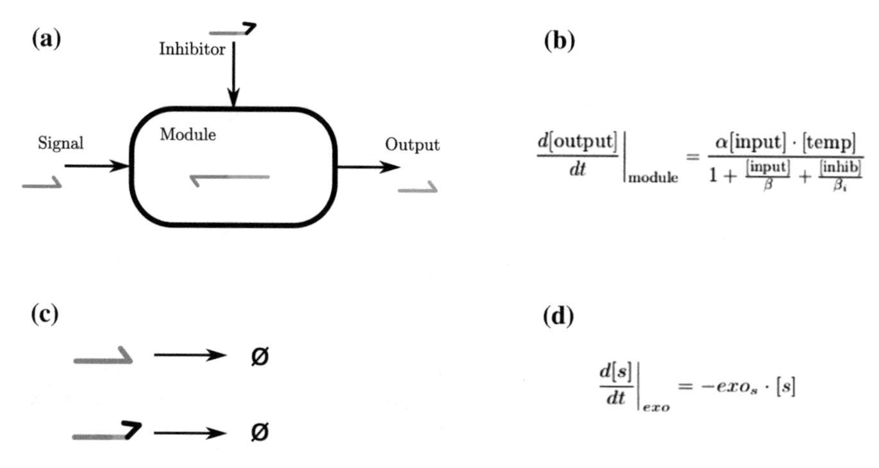

Fig. 5.20 **a** Black box representation of a template. **b** The general form of the transfer function for this box, with α the kinetic constant associated with the module, [.] the concentration of DNA species, β and β_i the saturation parameters of the input and inhibitor, respectively. **c** The reactions involving the exonuclease: Each unprotected species (anything but templates) is degraded over time. **d** The contribution of the exonuclease to the equation describing the evolution of the species s over time. exo_s, the activity of the exonuclease with respect to s, usually depends on the length of s. In general, this means that two values are possible, one for signals and one for inhibitors (slightly longer)

than enzyme-based reactions. This yields the transfer function in Fig. 5.20b. for each module. The whole complete set of equations can thus be easily built from the graph representation of a system by summing over templates generating a given species, with the addition of a first-order approximation of the exonuclease activity $-\text{exo}_s[s](t)$.

This model of the DNA PEN Toolbox was first introduced in Padirac et al.'s work on the bistable switch [16]. Because of the simplicity of the equations, large systems can be simulated extremely fast. This property was leveraged to evolve complex DNA PEN Toolbox systems, requiring thousands of separate evaluations. In this process, interesting patterns were discovered by the algorithm [3] (Fig. 5.21).

Applying the principles of this model, the behavior of an autocatalytic template can be written as a single equation:

$$\frac{d[s]}{dt}(t) = \frac{\alpha \cdot [temp][s](t)}{1 + \frac{[s](t)}{\beta}} - exo_s \cdot [s](t) \quad (5.6)$$

5.4.2 ERNe Details

The molecular networking of the DNA PEN Toolbox outlined in Sect. 5.4.1 rests on a combination of activatory or inhibitory chemical processes linking chemical

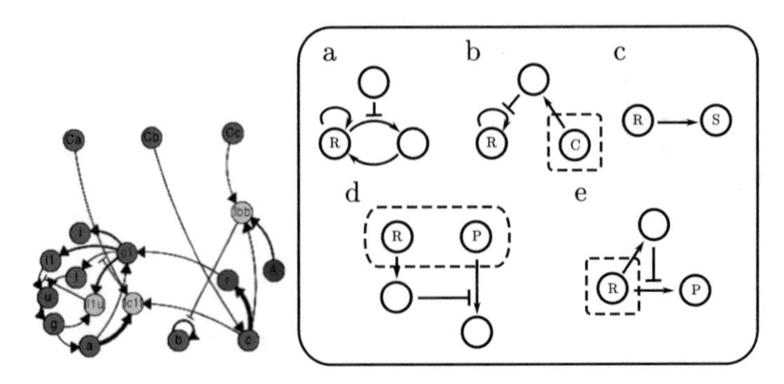

Fig. 5.21 Left: a large evolved system sensing its environment (three nodes at the top) and computing a new internal state. Right: various recurring patterns generated during repeated runs. One may consider them "standard" subroutines that are combined together to make larger systems

compounds considered as the nodes of a network. In this sense, they bear a structural resemblance to neural networks. Therefore, we apprehend that it might be possible to adapt modern Artificial Neural Networks (ANN) construction strategies to in silico evolution of biochemical systems, with a potentially large improvement in search efficiency. We use the extension of NEAT (Sect. 3.1.1) in this direction.

The main difference between ANN and reaction networks is the addition of inhibition links and biochemical parameters. Thus, ERNe encoding allows representation of inhibition and has added parameters. In addition, mutation and crossover operators are also modified from the original NEAT algorithm. In this section, the algorithm is described in detail, reviewing the encoding, the genetic operators for mutation and crossover, and the speciation process.

Our genome consists of sequence genes and template genes. Different from NEAT's node genes, each sequence gene can represent either a signal sequence or an inhibiting sequence in the system, and consists of a name, a kinetic parameter, and an initial concentration. Each template gene specifies the from-node, the to-node, the template concentration, an enabled bit that indicates whether or not the template is enabled, and an innovation number that uniquely identifies each template in the system.

The innovation number plays an important role in the implementation of the crossover and speciation, and is set for template genes based on the following rule. During the evolution, whenever a template is added to the system, we check if that specific link exists in the evolution history; in this case, it takes the original link's innovation number. Otherwise, the next available innovation number will be assigned to the template gene. To make this historical marking effective, a naming mechanism for newly added sequences should be carefully designed. Ideally, in different systems, sequences with the same name should carry the same role. It is, however, almost

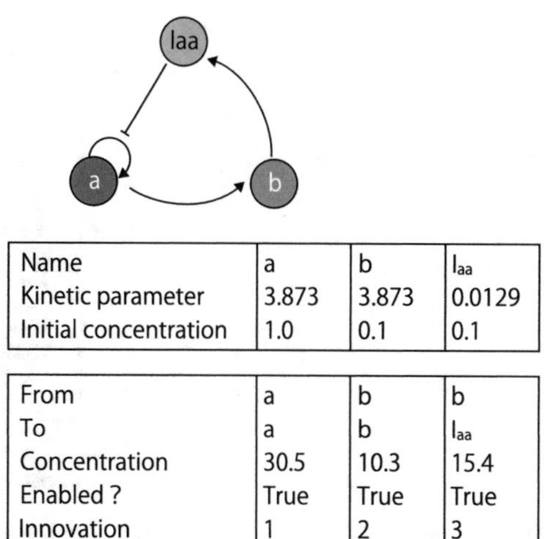

Sequence (Node) genes

Name	a	b	I_{aa}
Kinetic parameter	3.873	3.873	0.0129
Initial concentration	1.0	0.1	0.1

Template (Connection) genes

From	a	b	b
To	a	b	I_{aa}
Concentration	30.5	10.3	15.4
Enabled ?	True	True	True
Innovation	1	2	3

Fig. 5.22 A graph representation and corresponding ERNe encoding of the Oligator. Nodes represent sequences while arrows represent templates; bar-headed arrows represent inhibition. a and b are signal sequences, whereas the green node Iaa is an inhibiting sequence; it inhibits the template a→a (i.e., the self-activation of a). The system has three sequences and three templates. Thus, there are three node genes and three connection genes in its ERNe encoding

impossible to keep that ideal condition as later mutations might change the role of any sequence. However, a simple naming mechanism can partly deal with this matter. We use a list to map node names with the way they are created; for example, an entry A→A to B shows that whenever a new node is added in the middle of the template A→A, it must be named B. An example of genetic encoding describing the Oligator[9] is shown in Fig. 5.22.

5.4.3 Genetic Operations in ERNe

Mutations can be applied to change both the parameters and the network structure. We have the following mutation operators: parameter only, disable template, switch template, add sequence, add activation, and add inhibition. Their relationship and effects on the network are shown in Fig. 5.23.

In parameter mutation, every parameter has a probability to be mutated to a new random value calculated as follows:

$$newValue = oldValue \times (1 + f_1 \times rand_1) + f_2 \times rand_2,$$

[9] A robust DNA-enzyme oscillator.

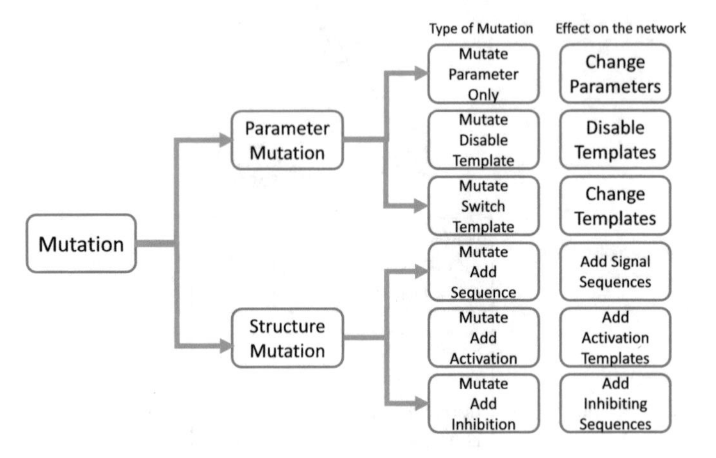

Fig. 5.23 Different types of mutation used in ERNe. Note that although mutate disable template and mutate switch template belong to parameter mutation, they actually change the structure as well

where f_1 and f_2 are fixed (set to 0.2 and 2.0, respectively, in the following experiments), and *rand* are standard normal deviates. To maintain the biochemical relevance, the available parameter range is bounded within physically reasonable values, and two further checks are performed on the mutated concentration: First, if that value is below zero, a switch template mutation happens. Indeed, negative concentrations do not exist, so we physically interpret this change as a switch to an inhibitory template. An example of switch template mutation is shown in Fig. 5.24a. In the figure, switch template mutation creates the structure (a) in which the template b→a is disabled, an inhibition node Iaa is added, and then a template b→Iaa is added with a new innovation number.

Assuming template b→a's concentration is mutated to a negative value, the changes are applied to the system, as described by Algorithm 5.1. Second, if the new value of a template concentration is above zero but below a threshold (here set to 1.0 [nM][10]), it has a probability to be disabled through a disable template mutation. These mutations are based on the idea that once a connection has no meaning (template concentration close to zero), it should be removed (disabled) or even changed to a reverse polarity. Also, because each template brings some variables to the ODE (Ordinary Differential Equation) model, these steps help to maintain structure as simple as possible by enabling the removal of templates with a very low concentration, whose relevance to the target function is presumably low.

In Add Sequence Mutation, a new signal sequence is added to the system. There are two ways to perform this type of mutation, as described in Fig. 5.24. The first

[10]nM is a unit of concentration in chemistry, where 1 [nM] $= 1 \times 10^{-9}$ [M].

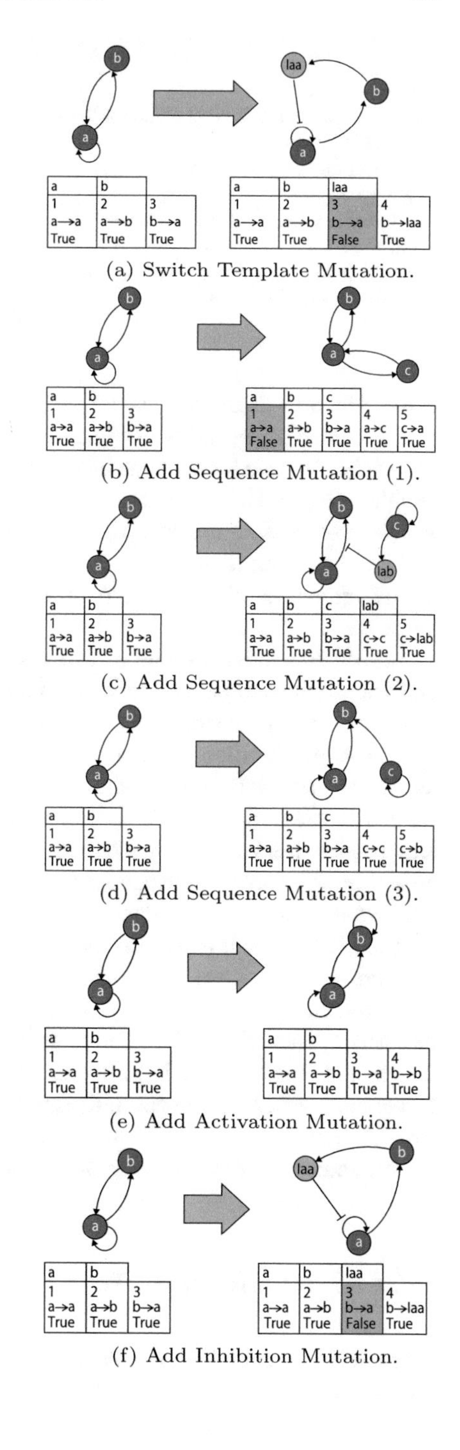

(a) Switch Template Mutation.

(b) Add Sequence Mutation (1).

(c) Add Sequence Mutation (2).

(d) Add Sequence Mutation (3).

(e) Add Activation Mutation.

(f) Add Inhibition Mutation.

Fig. 5.24 Several kinds of mutation in ERNe. An original structure is in the left

Algorithm 5.1 Switch Template Mutation

```
disable template b→a
if no existing template that creates a then
    add the template a→a
    select the template a→a
else
    select any template that creates a
end if
add the inhibiting sequence Ixa that inhibits the template x→a selected previously
add the template b→Ixa
```

is to select an existing template, split it, and place the new sequence in the middle (Fig. 5.24b). The second is to add a new sequence with a self-activatory connection (remember that nodes without incoming link are forbidden) to inhibit an existing template (Fig. 5.24c) or to activate an existing sequence (Fig. 5.24d). In the figure, Add Sequence Mutation creates the structure (b) in which a sequence c added in the middle of existing template a→a, the structure (c) in which a sequence c added to inhibit template a→b, and the structure (d) in which a sequence c is added to activate sequence b.

In Add Activation Mutation, two unconnected sequences are chosen randomly and a new template is added between them (Fig. 5.24e). In the figure, Add Activation Mutation creates the structure (vi) in which the template b→b is added.

The Add Inhibition Mutation's example is described in Fig. 5.24f. A random template is selected (a→a in the example) and the corresponding inhibition node (Iaa) is created. Then, the mutation selects a random start node (b) and adds the template b→Iaa, completing the module. A special case arises when there is already an activating template between sequences a and b; in this case, it is simply disabled (the mutation is then equivalent to a Switch Template Mutation).

The ERNe encoding and the use of innovation number make crossover straightforward. Using innovation numbers, the template genes in both parents are lined up, and then crossover techniques such as one-point and two-point crossover can be easily applied. Figure 5.25 shows an example of a two-point crossover operation, which leads to an interesting increase in loop size. We also need to decide how to create node genes for the child. Currently, we simply reconstruct it by reading the template genome in order of innovation number and select the parameters of the corresponding node in the parent that provided the connection.

5.4.4 Speciation in ERNe

When evolving complex systems, it is important to protect topological innovations. Indeed, smaller structures tend to optimize faster than larger ones. Moreover, in

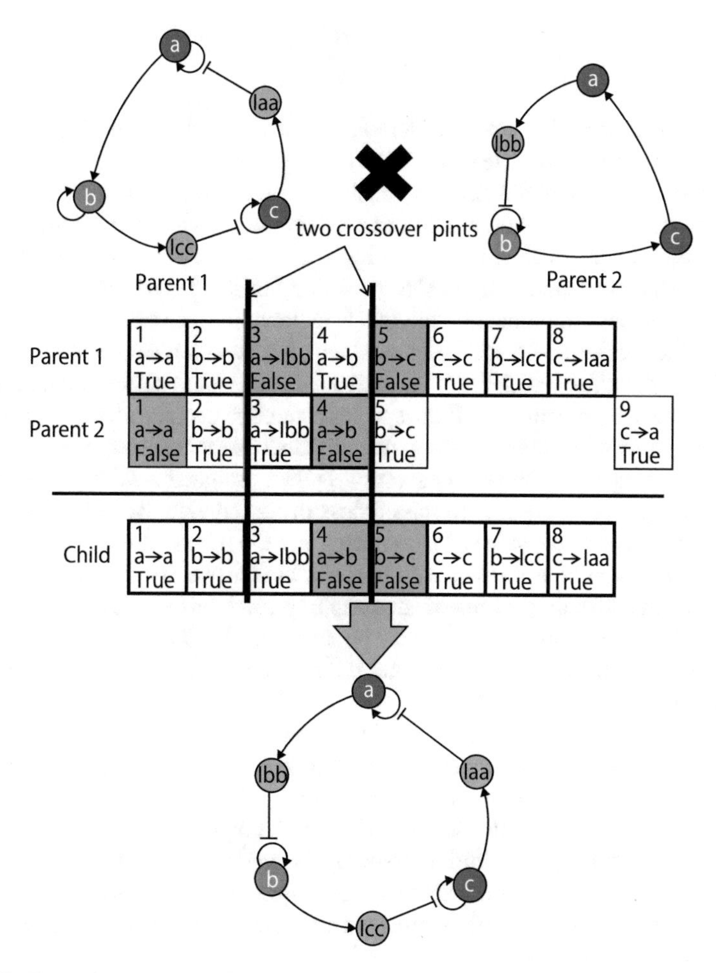

Fig. 5.25 Two-point crossover of the two individuals Parent 1 and Parent 2. The template genes are lined up in the order of innovation number. The red lines show the two crossover points. The child takes the genes number 1, 2, 5, 6, 7, 8 from Parent 1, and genes number 3, 4 from Parent 2. As a result, we have a new topology that has some parts of both parents

many cases, adding new nodes and connections to the system initially decreases the fitness. Thus, newly evolved structures rarely survive more than one generation, even though the innovations they introduce might be crucial toward solving the task in the end [19]. Therefore, we utilize the concept of species in which individuals with similar topologies are subgrouped in the same species. Consequently, individuals compete primarily within their own niches instead of with the whole population. This way, topological innovations are protected and get a chance to optimize. For each species, we call the first individual that discovers the species its representative. We use a topological compatibility distance δ calculated as follows:

$$\delta = \frac{M}{N},$$

where M the number of mismatched template genes between the two individuals and N the number of total template genes in the larger genome. If δ is below the speciation threshold δ_t, we say that the two individuals can be placed in the same species. We then use the following method to determine species for a newly created individual. We calculate its distance to all species' best individuals in the previous generation. The first match (having distance below the speciation threshold δ_t) decides its species. If no suitable species are found, all individuals in previous generations are considered. If no suitable species could still be found, then all species from the evolution history will be considered, in which case, the distance is calculated between the individual and the species' representative. If the species cannot be decided after all those steps, then a new species is created, having the individual as the representative.

Every species at the current generation is then assigned a different number of offspring in proportion to the average fitness of its individuals. Using the average fitness, individuals in the same species must share the fitness of their niche. In other words, a species is considered better if it has higher average fitness from all the individuals. In addition, we implement a capping mechanism that limits the growth of a species to at most 10% of the whole population. This prevents some temporally good species from growing too fast and taking over the population, leaving other species less chance to optimize.

Another problem introduced by speciation is the number of species at each generation. If we have too many species, each might not have enough space to evolve. In contrast, low number of species means low diversity. Actually, the number of species depends on the speciation threshold δ_t. With a high compatibility threshold, we tend to have less species. In our initial attempts, the performance of the run crucially depended on the value of this parameter. To bypass this issue, we implemented a dynamic adjustment of speciation threshold so that the number of species will move toward a specific target N_s:

$$\delta_t = \begin{cases} \delta_t + \epsilon & \text{if number of species } > N_s \\ \delta_t - \epsilon & \text{if number of species } < N_s, \end{cases}$$

where ϵ is called the modification step.

5.4.5 How ERNe Can Evolve GRNs

Our initial tests confirmed that ERNe could readily find already known oscillatory [13] or bistable [16] networks. Starting from a single node, it needs only a few generations to find these structures and optimize their parameters.

We first tried to find oscillating biochemical systems matching specific periodic functions: sine, rectangular, and sawtooth oscillations. The target outputs are generated with reasonable amplitude and period. We use a lexicographic fitness function that first checks the presence of oscillations and then calculates the fitness as follows:

$$fitness = \frac{1000}{MSE} \times limitcycle,$$

where *MSE* is the mean square error between the concentration of signal sequence a and the target function (the geometric MSE was used for the two discontinuous functions) and *limitcycle* evaluates the proximity a limit cycle by comparing the amplitude of the first and last peaks of the signal sequence a's concentration in the looking interval (set to 300–1500 min in our experiments).

There has been no research showing or giving hints about how such precisely shaped oscillations could be obtained. Thus, our initial attempts were to perform parameter optimization on a recently reported and controllable oscillatory structure built out of the DNA PEN Toolbox, the Oligator. The method used was standard DE (see Sect. 2.1.2) with population size of 40 and generation number of 1000. Our best attempts to match the sine, rectangular, and sawtooth oscillations which resulted in MSE of 31.6079, 34.3577, and 25.3381, respectively (time courses shown in Fig. 5.26), were unsatisfactory. It appears that, without changes in topological structure, the optimizer cannot simultaneously meet the required oscillation in terms of period, amplitude, and shapes. Thus, we believe these targeted time traces could only be generated by more complex structures that are impossible or at least very difficult to predict in advance. Later in this section, our discovered systems are shown to match these oscillations much better. Their structures are, indeed, much more complex, and in many cases, unrelated to the oscillator.

All our runs use the parameter settings shown in Table 5.7. These parameter settings were decided upon after some tuning on different problems and might not be the best set. However, we found it to be efficient for the benchmark runs. For elitism, the best individual of each species is copied to the next generation unchanged. All the runs started from an initial individual that contains only one signal sequence (and hence its self-activatory connection).

In order to evaluate which features of ERNe are the most important, we performed an experiment using one of the hardest problems, i.e., rectangular oscillation. This experiment compared four algorithms, each of which was run 50 times. The first two algorithms were ERNe with and without crossover. The third used crossover, but without speciation (like the simple GA that has been used widely in related approaches). The last test is ERNe with the original NEAT crossover technique.

The detailed result is given in Fig. 5.27, showing the performance over generations of all the runs. In addition, we defined a satisfactory fitness of 500, with which outputs of some example systems are shown in Fig. 5.28. Then, the average generations required to discover a solution with satisfactory fitness are shown in Table 5.8, and their statistical t-test results are shown in Table 5.9.

These p values indicate that there are significant differences between full ERNe and others' average numbers of generations required to achieve a satisfactory solution. It could be observed from Fig. 5.27 that the full ERNe with crossover plus

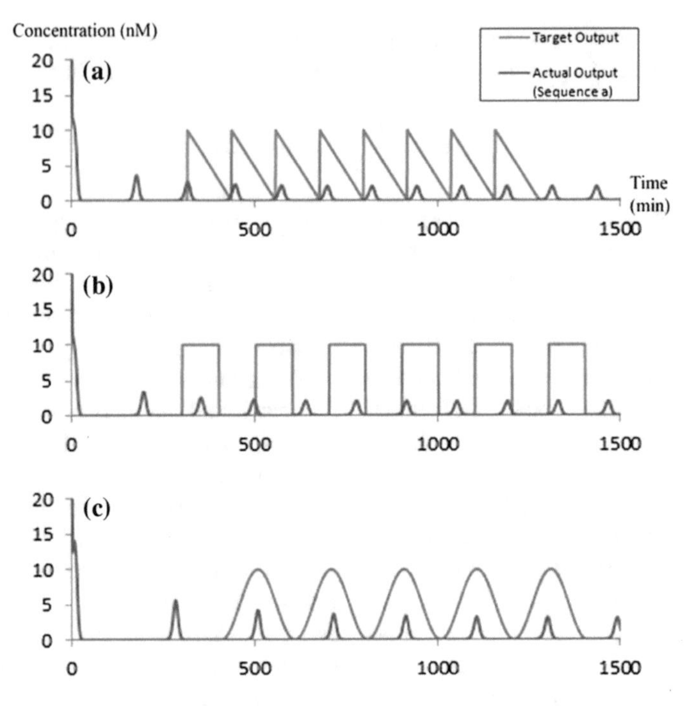

Fig. 5.26 Best time traces obtained for **a** sine, **b** rectangular, and **c** sawtooth oscillations by applying standard DE on optimizing the Oligator's parameters

speciation is the best. Performance drops significantly in the runs without speciation. Original NEAT crossover performed poorly, even worse than when no crossover is involved.

However, with ERNe one-point crossover applied, a solution with satisfactory fitness was found about 50% faster than without crossover on average. These results suggest that crossover, if used properly, plays an important role to speed up the evolution, but those good solutions can also be found without crossover, with the cost of larger runs. Moreover, the fact that the original NEAT crossover, which was efficient in evolving ANN, is disruptive in this problem suggests that differences between ANN and biochemical reaction networks are significant and should not be overlooked.

As the performance result suggests, ERNe with crossover showed highest performance. Thus, we use ERNe with crossover for the rest of the experiments. For each of the three oscillation shapes, the algorithm was run for 50 times. Importantly, the three problems were addressed with the exact same settings, shown in Table 5.7.

Figure 5.29 shows a typical result for the sine oscillation problem. The observed output and the target output are matching almost perfectly. The structure is based on an extended version of the Oligator structure (Fig. 5.22) with a longer delay line in

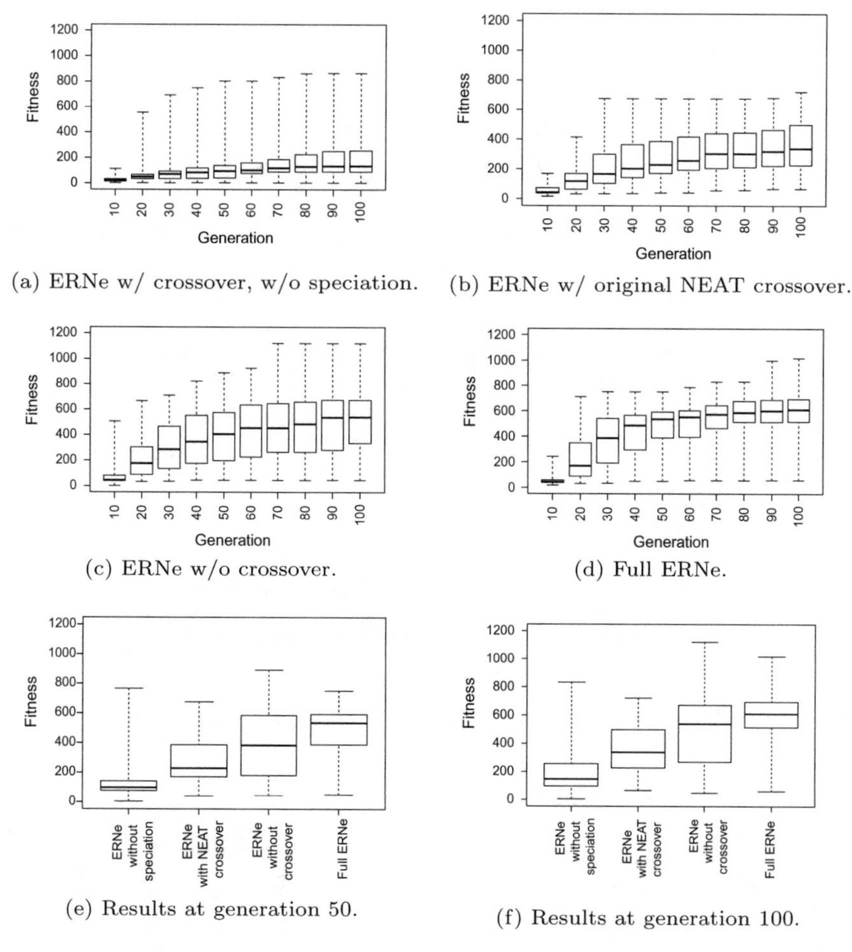

Fig. 5.27 Performance results for rectangular oscillation problem. Best fitness values over generations are displayed in box-plot for **a** ERNe without speciation, **b** ERNe with original NEAT crossover, **c** ERNe without crossover, **d** full ERNe, **e** snapshot of all the runs at generation 50, and **f** snapshot of all the runs at generation 100

the negative feedback loop. Side feed-forward loops are grafted to this structure and probably serve to fine-tune the shape of the time trace.

Figure 5.30 shows results for the rectangular oscillation. As expected, the fitness value is not as good as that observed previously for continuous functions. Interestingly, most runs (40/50) converged toward an unreported and unrelated oscillatory topology involving three self-activating sequences linked together by three inhibition

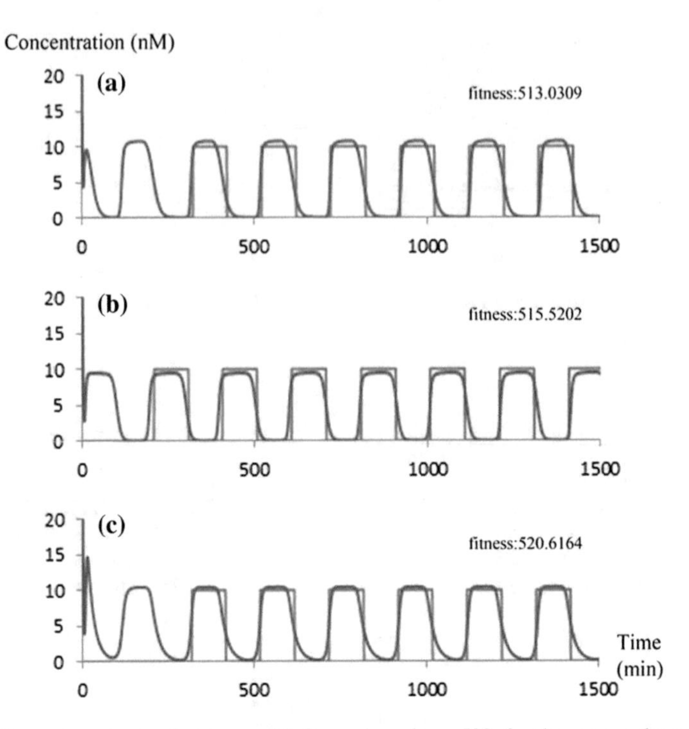

Fig. 5.28 Example outputs of systems with fitness just above 500, for the rectangular oscillation problem

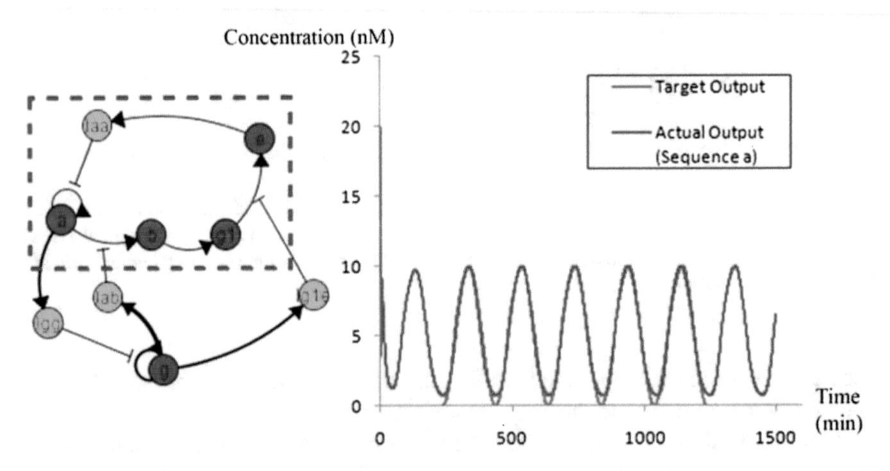

Fig. 5.29 Typical solution for the sine oscillation problem, with structures based on an elongated Oligator (the structure in the red dotted rectangle), and their corresponding output

Table 5.7 Parameters used in experiments

General parameters	
Population size	200
Number of generations	100
Selection method	Tournament (size 5)
Crossover technique	One-point crossover
Speciation parameters	
Preferred number of species	10
Starting speciation threshold	0.6
Minimum speciation threshold	0.1
Speciation modification step ϵ	0.03
Crossover and mutation parameters	
P(Mutation only)	0.5
P(Interspecies mating)	0.01
P(Mutation after crossover)	0.75
Mutation parameters	
P(Parameter only)	0.9
P(Single gene mutation)	0.8
P(Structure-add node)	0.2
P(Structure-add activation)	0.2

Table 5.8 Generations required to achieve fitness above 500 for the rectangular oscillation problem

ERNe without speciation	145.6 ± 22.1
ERNe with NEAT crossover	125.2 ± 45.9
ERNe without crossover	84.2 ± 61.1
Full ERNe	59.6 ± 50.7

Table 5.9 p values of t-distribution calculated from Table 5.8

Full ERNe versus ERNe without speciation	1.48E-09
Full ERNe versus ERNe with NEAT crossover	2.59E-5
Full ERNe versus ERNe without crossover	0.0310

reactions. We called this topology the switch oscillator in reference to its tendency to produce fast switching separated by flat plateaus (Fig. 5.30(i)). Within these 40 runs, 34 were found to contain the extended variant where signal sequence a (the output node that we are tracking) is not self-activating but serves as an intermediate compound (Fig. 5.30(ii)) while the remaining 6 had an additional counter-rotating

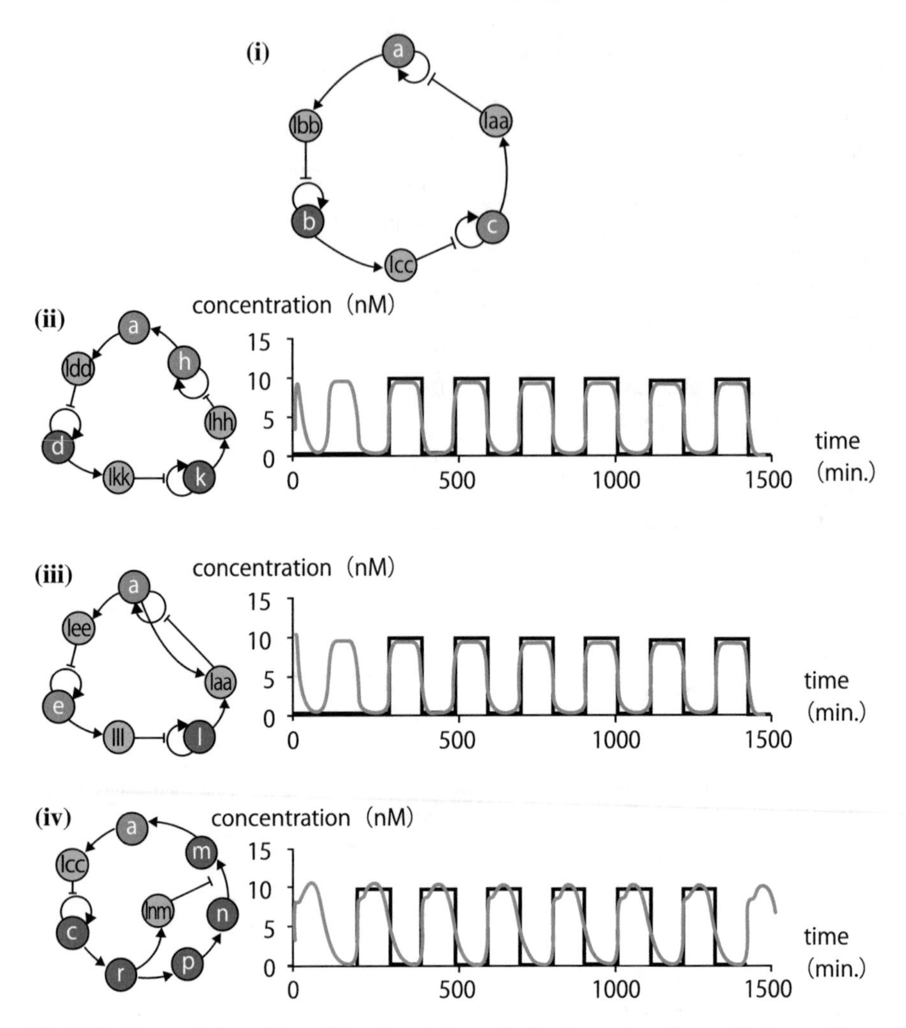

Fig. 5.30 New switch oscillator (i) and typical results for the rectangular oscillation problem; 40/50 of the runs converged to a topology derived from (i) where three self-activating sequences linked together by three inhibition reactions and including either an extension (ii, 34 runs) or a self-inhibition (iii, 6 runs) located at the stage. The rest (10/50 runs) fell into the long Oligator category (iv) and did not match well

activation also originating from a (Fig. 5.30(iii)). This robust convergence to precise topological features suggests that the framework is effective in exploring the search space. The remaining 20% of the runs fell into the long Oligator family (Fig. 5.30(iv)). Even after intensive tuning, this class of oscillators was unable to accurately match the constant parts of the target function and produced lower fitness.

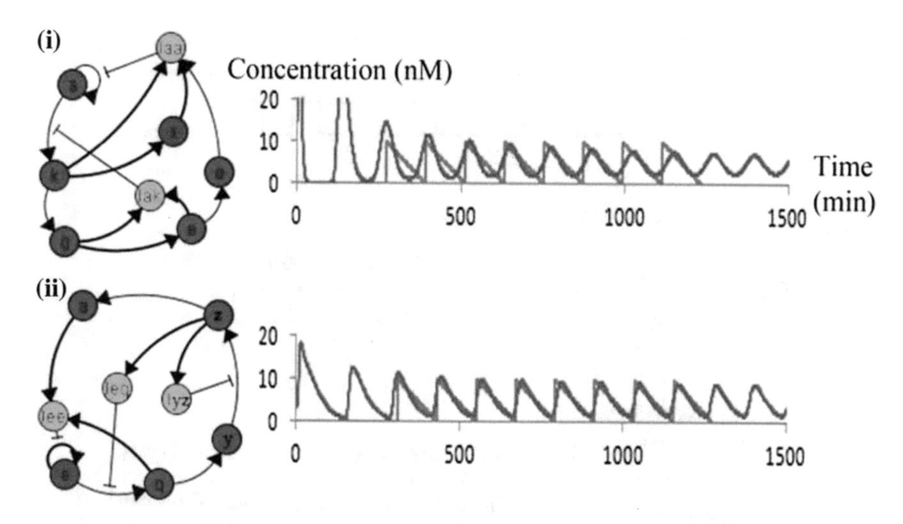

Fig. 5.31 Typical structures and their sample outputs for the sawtooth oscillation problem. All the runs converged to the topology of elongated Oligator, in which 7/50 topologies have sequence a as self-activating sequence (i), and 43/50 have sequence a as the last of the negative feedback loop (ii). The latter provides better solutions

Figure 5.31 shows the results for sawtooth oscillations. As expected given the relaxation form of the target function, all runs converged to a topology based again on the long Oligator (which is a relaxation oscillator). We could again observe the repetitive convergence to precise topological details, such as the target sequence being either self-activating (7/50 runs with average MSE of 5.2004 ± 1.3651, Fig. 5.31(i)) or the last one of the negative feedback loop (43/50, producing significantly better individuals with average MSE of 1.9422 ± 0.5343, Fig. 5.31(ii)). In this case, also, side branches serve to fine-tune the shape of the temporal evolution.

One example of the evolution to match the rectangular oscillation is shown in Fig. 5.32. The initial network (i, generation 0) mutated to (ii, generation 1) by adding a new node d in the middle of the self-activation $a \to a$, with no improvement in fitness (no oscillation means zero fitness). Later, the activation $a \to d$ was switched to inhibition, creating the Oligator motif (iii, generation 2). This structure is compatible with oscillation; however, the parameters are not yet optimized. Later, it mutated to a long Oligator (iv, generation 3)—which is a more robust oscillator—by adding a new node h. The system (iv) optimized its parameter for several generations and could reach a strong oscillation with a fitness of 40.8673 at generation 8. At generation 9, another new node k is added to the system to elongate the feedback loop of the

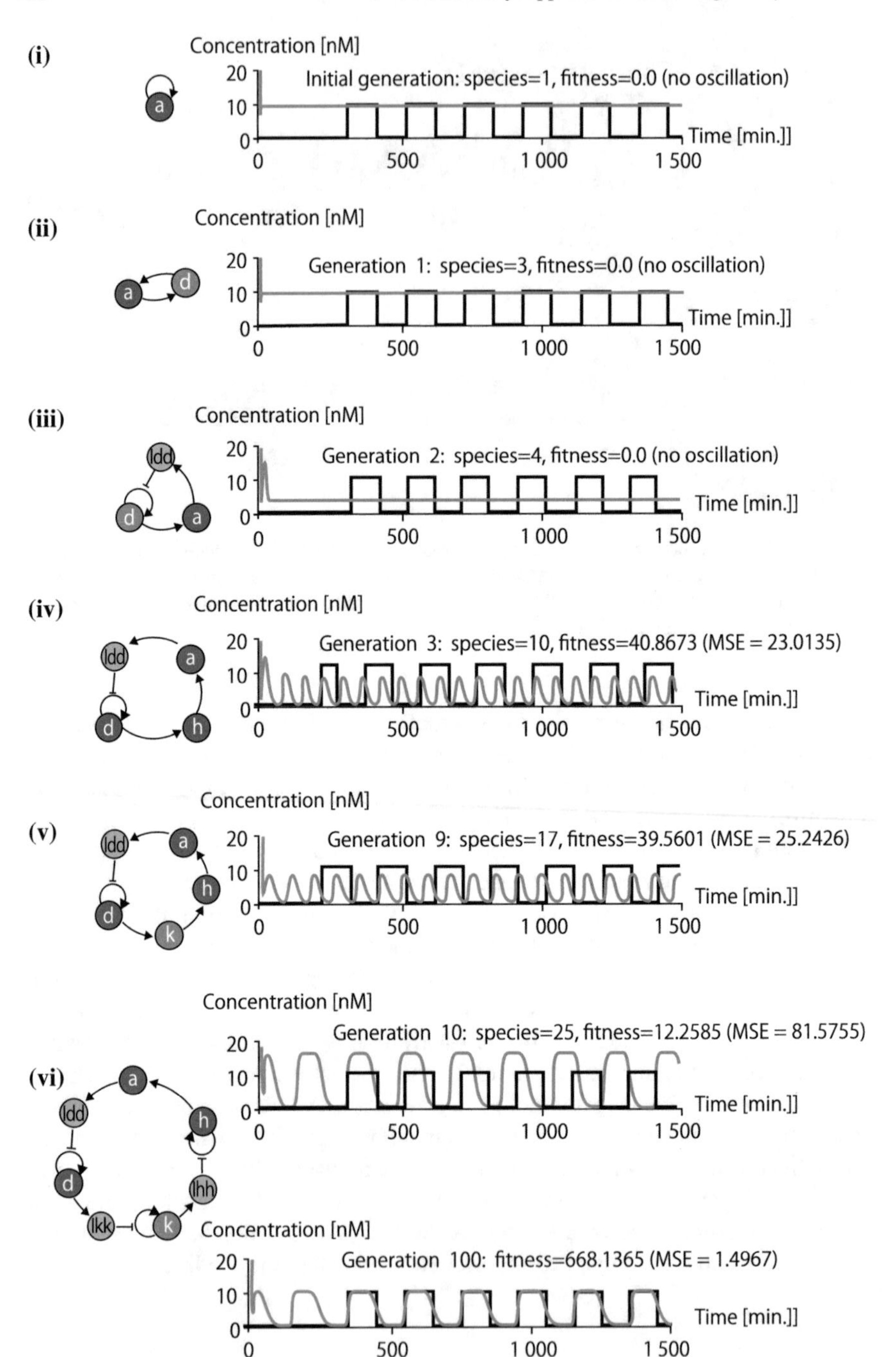

Fig. 5.32 Evolution of a network for the rectangular oscillation. Only the changes to the topology are shown. There were, in fact, many parameter optimizations during the evolution

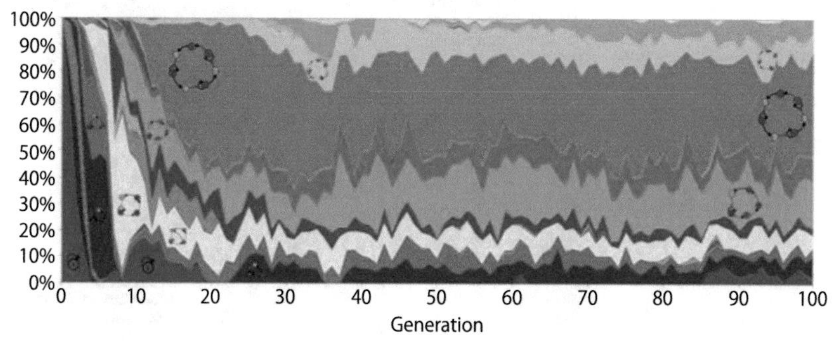

Fig. 5.33 Species visualization for a sample run of the rectangular oscillation problem

system and results in system (v). This mutation decreased the fitness slightly. The final structure which is a derivative of the switch oscillator (Fig. 5.30(i)) could be obtained at generation 10, when two switch mutations happened at the same time to the activation d→k and k→h. Interestingly, a huge fitness drop (to 12.2585) was experienced with the new structure. Although the oscillation it produces is more similar to rectangular oscillation, the mismatch in both amplitude and period led to high MSE. Thanks to the speciation, this structure could be kept in the population, optimized, and finally became the best solution with fitness of 668.1365. Here, we only focus on the mutations that changed the structure. There were, in fact, many parameter mutations coming along that also played important roles.

For this specific run, also, the speciation process is depicted in Fig. 5.33, where the generations are shown from the left to right, with the species depicted vertically for each generation. The height of each species is proportional to its population in that generation. It can be clearly seen that the switch oscillator species appeared early with a small population then became dominating later. The long Oligator species, on the other hand, always performed well and occupied a stable portion of the population. It is also interesting that most initial species disappeared before generation 10, indicating that simple structures are not suitable to solve this complex problem.

5.4.6 Application of ERNe

Consider a simple Boolean experiment with ERNe. First, we synthesize a reaction network that solves the XOR problem, which is often employed to test neural networks. As shown in Fig. 5.34, the task is given as a problem involving two-bit input vectors (i.e., 00, 11, 01, 10) yielding one-bit output (i.e., 0, 0, 1, 1). To solve this task, we need two appropriate nodes as input and one node as output. We run the network so that the concentration of an input node is 0 [nM] for a binary 0 and 25 [nM] for a binary 1, and then train the network so that the concentration of the output node

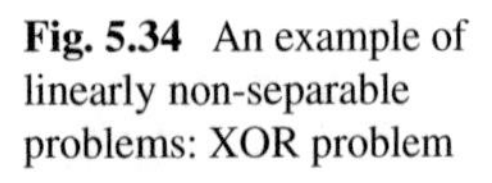

Fig. 5.34 An example of linearly non-separable problems: XOR problem

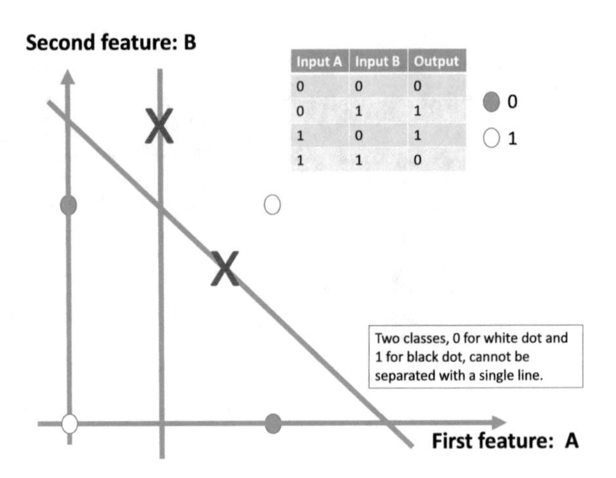

Table 5.10 ERNe parameters for XOR and inverted pendulum problem

Parameter	Value
Population size	1000
Max. generation	100
Selection	tournament (size 20)
Crossover	One-point crossover
Sharing parameter	Value
#. of species	20
Threshold of sharing to start	0.6
Minimum threshold of sharing	0.1
Sharing modification step	0.03
Probabilities for genetic operators	Value
Crossover + mutation prob.	0.8
Crossover between species	0.01
Mutation after crossover prob.	0.75
Mutation prob. (without crossover)	0.5
Mutation prob. (single gene)	0.8
Node addition prob.	0.1
Activation prob.	0.4

approaches the output value in Fig. 5.34 after some time has passed and the reaction has stabilized.

Specifically, we convert the output concentration to a binary value by considering an appropriate threshold (in this case, 200 [nM]). Figure 5.35 shows the search results for ERNe working on the XOR problem. This figure shows the fitness value for each generation (the error when converted to the ideal concentration for the output value such that larger values are better), as well as the networks obtained at various points.

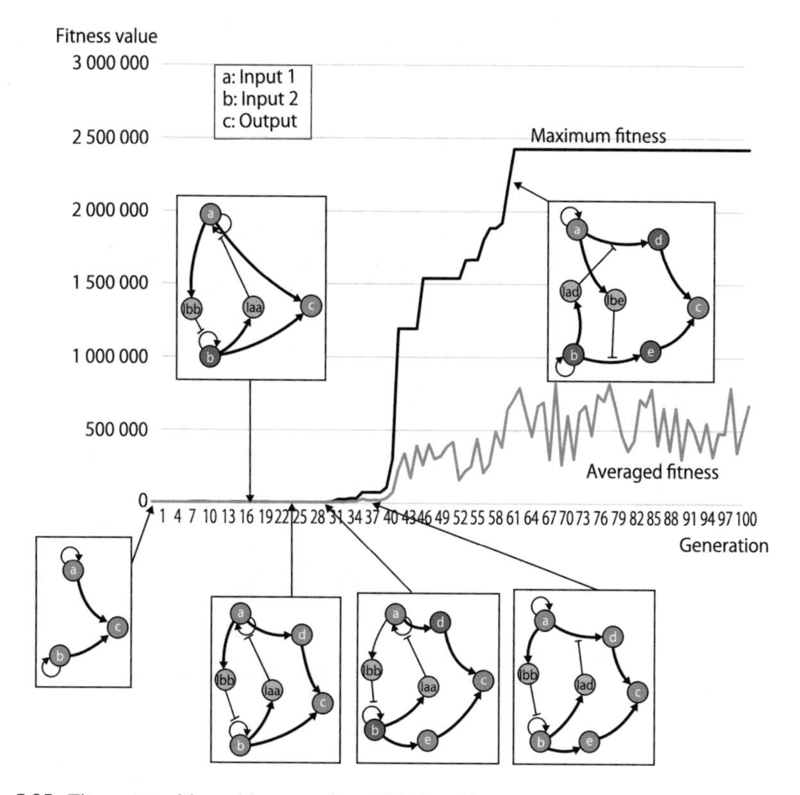

Fig. 5.35 Fitness transition with generations: XOR problem

Table 4.1 shows the ERNe parameters used in this experiment. The concentrations for *a* and *b* are used for the XOR input, while the concentration for *c* is the output. Here, we see that ERNe searches for (and finds) the structure accurately. In this case, the circuit for the correct answer was found after 61 generations. Figure 5.36 shows how the reactions for this network (the network evolved for the XOR problem) progress over time. The dotted line in the figure is the predetermined threshold (200 [nM]) for classifying the output as either 0 or 1. As can be seen in the figure, once the reaction reaches a steady state the concentration of the output node *c* behaves as an exclusive logical OR.

Next, we try to solve the problem of controlling an inverted pendulum (Fig. 5.37). An inverted pendulum is an unstable structure in which a pendulum is turned upside down, except that here we will consider a bar pendulum rather than a string. An example of this problem is the game of trying to balance an umbrella in the

Fig. 5.36 Network reactions (XOR problem)

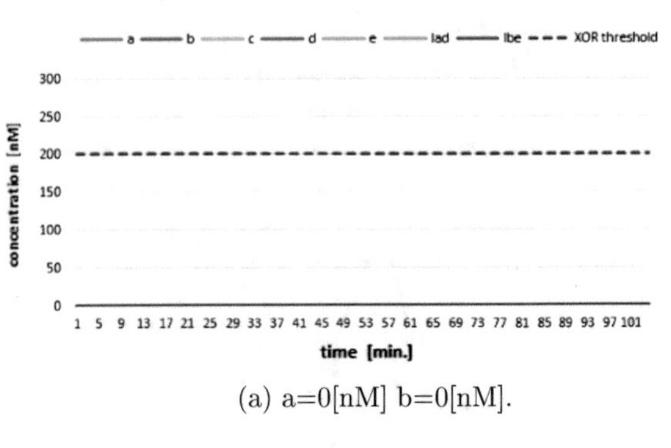

(a) a=0[nM] b=0[nM].

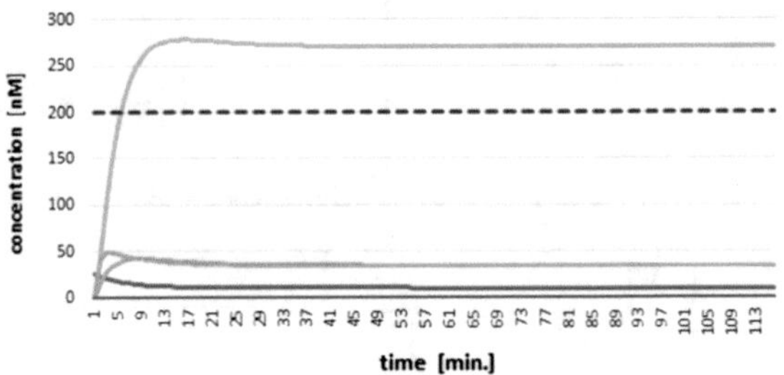

(b) a=0[nM] b=25[nM].

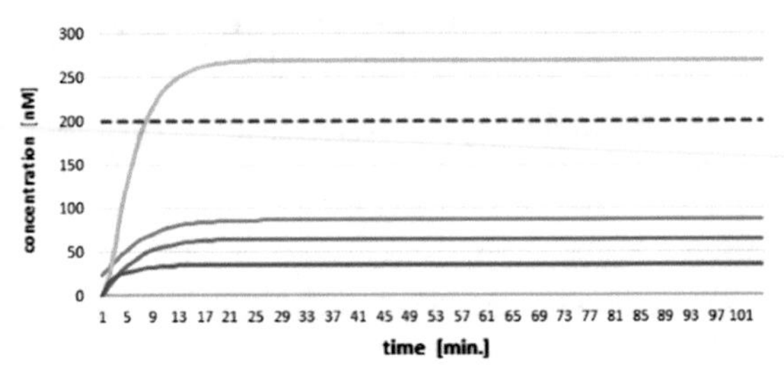

(c) a=25[nM] b=0[nM].

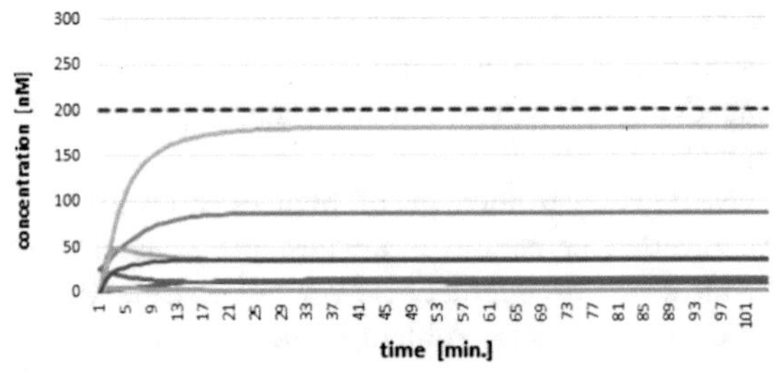

(d) a=25[nM] b=25[nM].

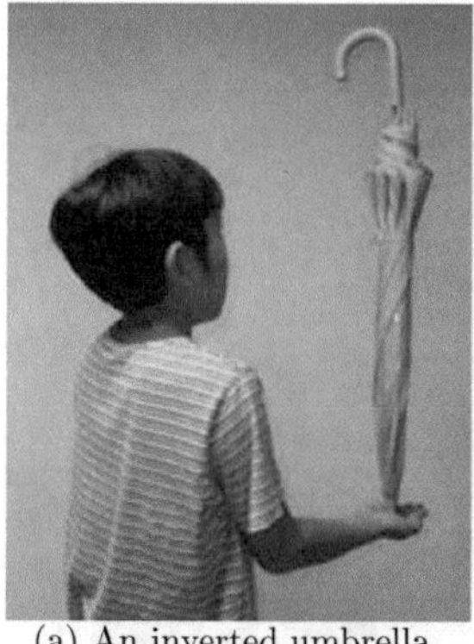
(a) An inverted umbrella.

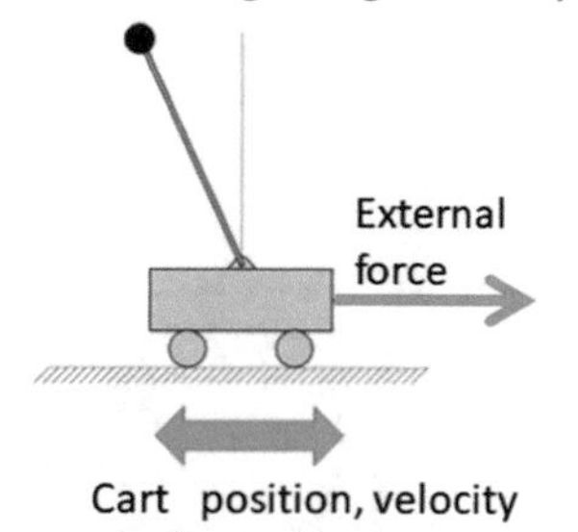

(b) Model of an inverted pendulum.

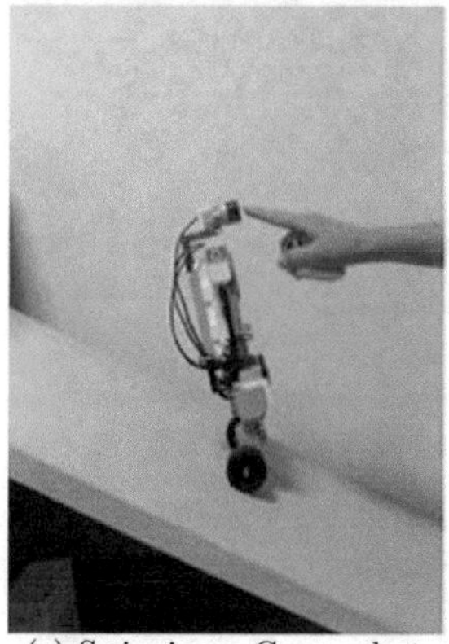
(c) Swinging a Gyro robot.

(d) Segway-type humanoid robot.

Fig. 5.37 Inverted pendulum problem

palm of your hand (Fig. 5.37a). To prevent the umbrella from falling, you need to move your palm (and thus the fulcrum of the pendulum) appropriately (Fig. 5.37b). In other words, an inverted pendulum is a basic example of feedback control (see other robot examples shown in Fig. 5.37c, d). We trained ERNe to control an inverted pendulum. Here, we conducted training using ten different example training patterns with random initial positions (pendulum angles). For the fitness value derivation, we look at the total time (in seconds) that the pendulum remains in a stable position (between 54° and 126°, so larger values (i.e., longer times) naturally indicate better fitness. The maximum execution time for each of the training examples was 10,000 s. Figure 5.38 shows the execution process. For this search, we used the parameters in Table 5.10. Interpretations for the nodes in the network are shown in the figure. For example, the concentration of a represents the input for the position of the cart, and the concentration of e is the output, representing the external force applied to the cart. In this case, a circuit that maintains perfect balance was found after 40 generations. In other words, the fitness value became $10 \times 10{,}000 = 100{,}000$.

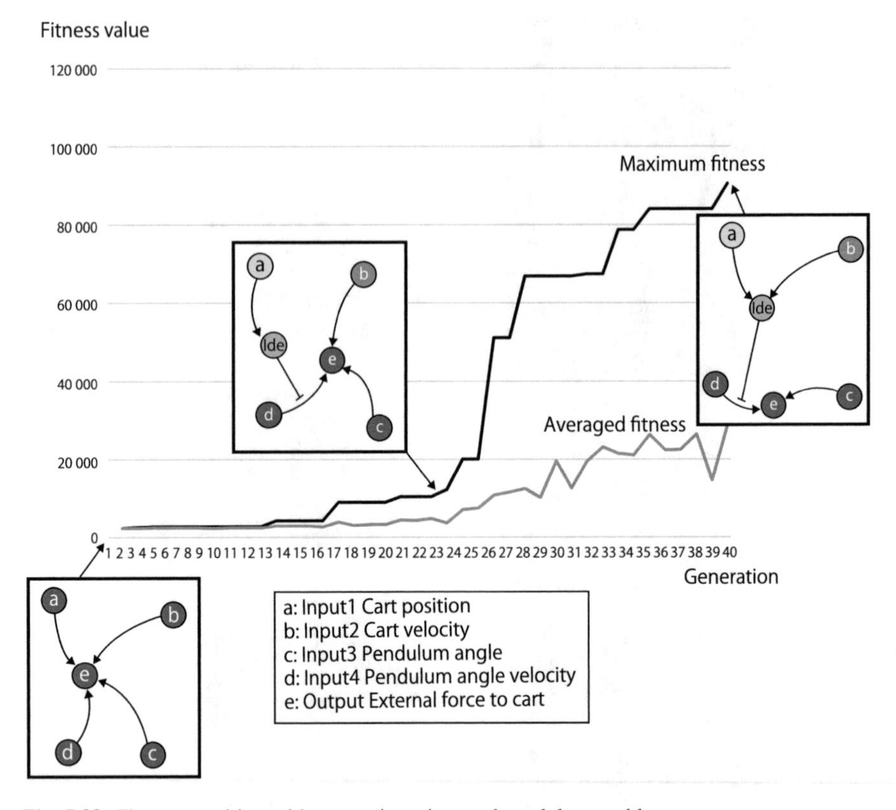

Fig. 5.38 Fitness transition with generations: inverted pendulum problem

Next, we actually verified the circuit obtained by ERNe in vitro. Figure 5.39 shows one of the results of using ERNe to design genetic circuits to implement numerical functions (such as log, Gaussian, and second power) and then actually conducting wet experiments using a DNA PEN Toolbox.[11] The details of the biochemistry are not important here, but note that the genetic reaction networks obtained by ERNe can have practical application in vitro.

This section has discussed practical examples of the ERNe mechanism. Figure 5.40 shows a framework for problem solving in artificial intelligence and robotic control using ERNe. The genetic circuits for solving a problem are first designed using ERNe, and then biochemical reactions are carried out in vitro. This method is different than DNA computing, which focused on genetic parallel processing because ERNe actively induces biochemical reactions by designing genetic circuits.

[11]The thermal cycler in the figure is a device that duplicates DNA fragments.

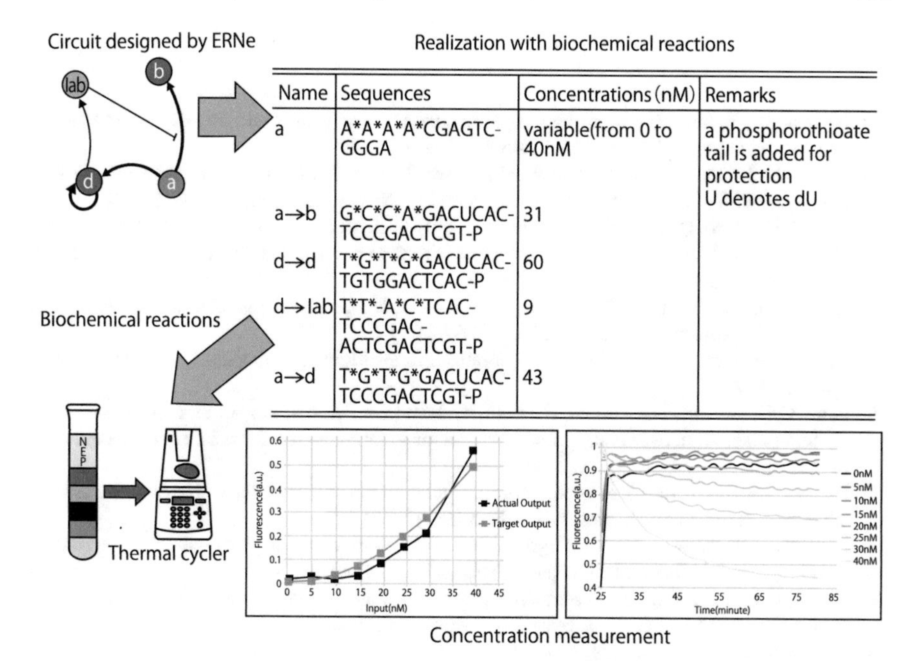

Name	Sequences	Concentrations (nM)	Remarks
a	A*A*A*A*CGAGTC-GGGA	variable(from 0 to 40nM	a phosphorothioate tail is added for protection U denotes dU
a→b	G*C*C*A*GACUCAC-TCCCGACTCGT-P	31	
d→d	T*G*T*G*GACUCAC-TGTGGACTCAC-P	60	
d→lab	T*T*-A*C*TCAC-TCCCGAC-ACTCGACTCGT-P	9	
a→d	T*G*T*G*GACUCAC-TCCCGACTCGT-P	43	

Fig. 5.39 In vitro verification with ERNe

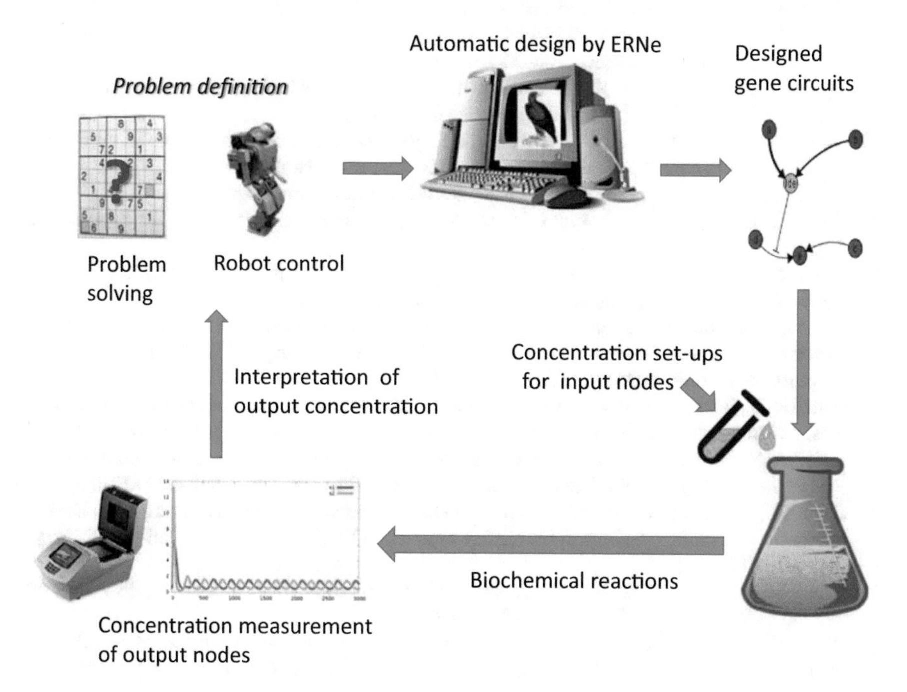

Fig. 5.40 ERNe for solving AI problems

References

1. Aldana, M., Balleza, E., Kauffman, S., Resendiz, O.: Robustness and evolvability in genetic regulatory networks. J. Theor. Biol. **245**(3), 433–448 (2006)
2. Alipanahi, B., Delong, A., Weirauch, W.T., Frey, B.J.: Predicting the sequence specificities of DNA- and RNA-binding proteins by deep learning. Nat. Biotechnol. **33**, 831–838 (2015)
3. Aubert, N., Dinh, Q.H., Hagiya, M., Iba, H., Fujii, T., Bredeche, N., Rondelez, Y.: Evolution of cheating DNA-based agents playing the game of rock-paper-scissors. In: Advances in Artificial Life, ECAL, vol. 12, pp. 1143–1150 (2013)
4. Chen, Y., Li, Y., Narayan, R., Subramanian, A., Xie, X.: Gene expression inference with deep learnings. Bioinformatics **32**(12), 1832–1839 (2016)
5. Cliff, D., Harvey, I., Husbands, P.: Explorations in evolutionary robotics. Adapt. Behavior **2**, 72–110 (2000)
6. Dinh, Q.H., Aubert, N., Noman, H., Fujii, T., Rondelez, Y., Iba, H.: An effective method for evolving reaction networks in synthetic biochemical systems. IEEE Trans. Evol. Comput. **18** (2015)
7. Iba, H., Noman, N. (eds.): Evolutionary Computation in Gene Regulatory Network Research. Wiley Series in Bioinformatics. Wiley, Hoboken (2016)
8. Inamura, T., Nakamura, Y.: An integrated model of imitation learning and symbol development based on Mimesis theory. Brain Neural Netw. **12**(1), 74–80 (2005)
9. Jin, Y., Guo, H., Meng, Y.: A hierarchical gene regulatory network for adaptive multi-robot pattern formation. IEEE Trans. Syst. Man Cybern. B Cybern. **42**(3), 805–816 (2012)
10. Kauffman, S.A.: The Origins of Order, Self-organization and Selection in Evolution. Oxford University Press, New York (1993)
11. Marbach, D., Costello, J.C., Kuffner, R., Vega, N., Prill, R.J., Camacho, D.M., Allison, K.R., The DREAM5 Consortium, Kellis, M., Collins, J.J., Stolovitzky, G.: Wisdom of crowds for robust gene network inference. Nat. Methods **9**(8), 796–804 (2012)
12. Mendes, P., Sha, W., Ye, K.: Artificial gene networks for objective comparison of analysis algorithms. Bioinformatics **19**(Suppl 2), 122–129 (2003)
13. Montagne, K., Plasson, R., Sakai, Y., Fujii, T., Rondelez, Y.: Programming an in vitro DNA oscillator using a molecular networking strategy. Mol. Syst. Biol. **7**, 466 (2011)
14. Nolfi, S., Floreano, D.: Evolutionary Robotics. MIT Press, Cambridge (2000)
15. Park, Y., Kellis, M.: Deep learning for regulatory genomics. Nat. Biotechnol. **33**(8), 825–826 (2015)
16. Padirac, A., Fujii, T., Rondelez, Y.: Bottom-up construction of in vitro switchable memories. Proc. Natl. Acad. Sci. (PNAS) **109**(47), E3212–E3220 (2012)
17. Palafox, L., Noman, N., Iba, H.: On the use of population based incremental learning to do reverse engineering on gene regulatory networks. In: Proceedings of the IEEE Congress on Evolutionary Computation (CEC), pp. 1–8 (2012)
18. Palafox, L., Noman, N., Iba, H.: Gene regulatory network reverse engineering using population based incremental learning and K-means. In: GECCO (Companion), pp. 1423–1424 (2012)
19. Palafox, L., Noman, N., Iba, H.: Reverse engineering of gene regulatory networks using dissipative particle swarm optimization. IEEE Trans. Evol. Comput. **17**(4), 577–587 (2013)
20. Palafox, L., Noman, N., Iba, H.: Extending population based incremental learning using Dirichlet processes. In: Proceedings of the IEEE Congress on Evolutionary Computation (CEC), pp. 1686–1693 (2016)
21. Qian, L., Winfree, E.: Scaling up digital circuit computation with DNA strand displacement cascades. Science **332**(6034), 1196–1201 (2011)
22. Zaier, R.: Motion generation of humanoid robot based on polynomials generated by recurrent neural network. In: Proceedings of the First Asia International Symposium on Mechatronics (2004)

Chapter 6
Conclusion

It struck me how little is known about these creatures of the rain forest, and how deeply satisfying it would be to spend months, years, the rest of my life in this place until I knew all the species by name and every detail of their lives.

(Edward O. Wilson, The Diversity of Life, Harvard University Press, 1992)

Abstract This chapter concludes the book with some summaries and ideas about evolution. As we have seen in the previous chapters, we can use evolutionary mechanisms mainly for designing desirable structures, not for the pure purpose of optimization. This is a common confusion of the fact of evolution being progress. Evolution is a constructer of better building blocks, not a optimizer of a simple solution. This is known as the concept of "Punctuated Equilibrium."

Keywords Optimization versus evolution · Punctuated equilibrium · Stephen. J. Gould · Niles Eldredge · Ultra-Darwinists · Selfish genes · Richard Dawkins

This book provided theoretical and practical knowledge about a methodology for EA with the integration of several AI techniques, i.e., deep learning and machine learning. As I mentioned in the Preface, EA has the following features:

- Parallelism
- Searchability
- Diversity

Among them, diversity is the most important, as we can see in biology. For example, there are many cases of coexistence with give-and-take relationships (Fig. 6.1a refers to coexistence of Vanderhorstia and snapping shrimp). Completely different types of organisms can act cooperatively and share work. There are also cases of social insects cooperating to build nests, such as ants and bees (Fig. 6.1b). In other words, referring to natural world mechanisms that cannot easily be understood, it may be

H. Iba, *Evolutionary Approach to Machine Learning and Deep Neural Networks*, https://doi.org/10.1007/978-981-13-0200-8_6

(a) Symbiosis of Vanderhorstia and snapping shrimp.

(b) Ants castle.

Fig. 6.1 Examples of biological diversities

possible to model organic activity and, by incorporating this into the evolutionary computation, bring about new discoveries.

Regarding the diversity, groups comprise various individuals. Under evolution, the high-performing members of the group tend to remain through greater productivity. However, this does not necessarily mean it is good to always assemble the high-performing elite members alone. Doing so can cause the group as a whole to decline if the environment changes. While similar descendants born from the elite group may perform well in the current environment, their ability to adapt to new situations is often lacking. This property (sensitivity to environmental change and noise) is equivalent to being "not robust" in engineering terms, and it should be avoided in designed systems where possible. Accordingly, it is better to allow the existence of poor and inferior individuals within the group (see also Fig. 1.8). While they are usually a burden, they can potentially become the savior of the group in the future. In this way, evolution is not necessarily the search for the optimal value, but rather the aim for robustness that enables a greater chance of survival in the environment.

In concluding this book, we shall discuss the key concept in EC, i.e., the relationship between optimization and evolution.

As evolutionary computation is modeled on evolution, it seems that many people consider "progress=optimization." Therefore, it is thought that evolution involves continual development. However, it is necessary to be aware that "evolution is not progress." We are not surviving by optimization.

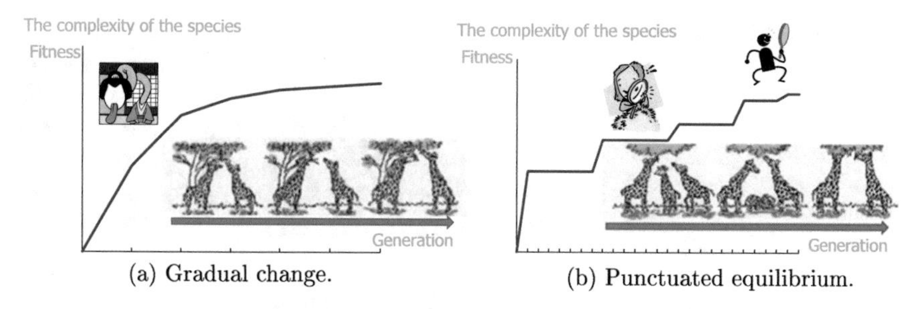

(a) Gradual change.

(b) Punctuated equilibrium.

Fig. 6.2 Evolution is not optimization

Let us apply the evolutionary computation and observe fitness behavior as usual. The greater the fitness, the better. In this case, you can see that there are the following two typical behaviors:

- Gradual change (Fig. 6.2a)
- Punctuated equilibrium change (Fig. 6.2b)

In the case of (a), gradual change (gradually progressing evolution), this is extremely suited to evolutionary computation. It may be too early to relax, though. However, this may be solved by using other simpler methods (e.g., hill-climbing). In many cases, we face the situation (b). This situation may be suited to evolutionary computation, but it is worth studying what kind of structures can be learned between the steps. In other words, the essence of the problem lies in the stagnant time. This may possibly improve if we just wait. For example, in an actual practical example, we can see that important components (i.e., substructures or useful building blocks) are acquired when climbing the steps of fitness. This kind of knowledge can be a major step toward improvements in learning.

In evolution as well, we can see that there is the same level of fitness (length of the giraffe's neck, complexity of species, etc.) and the generational relationships are as in (b). For example, we have found the fossils of organisms with short necks among the ancestors of giraffes, but we have not yet found any instances of giraffes with medium-sized neck lengths. This may be a case of "not looking hard enough" or the "bones and cartilage of the neck not easily forming fossils," but this does not sit well with us. In response to this, Stephen. J. Gould[1] and his colleague Niles Eldredge espoused the hypothesis of "Punctuated Equilibrium." Gould considered that "stagnation is also data" and that evolution occurred within the short period of time in which the species branched out, and between those times, there was a period of stagnation. Stagnation does not mean just not doing anything, and it makes up very interesting and precious evidence and can be considered to be interrupted by rarely occurring changes. During stagnation, neutral genetic recombination is frequently occurring, but due to some trigger, such as changes in the environment,

[1] An American paleontologist and evolutionary biologist (1941–2002). He wrote essays every month for "Natural History," an American science journal, and many collections of these became best sellers. He was an opponent of another evolutionary researcher, Richard Dawkins (see p. 6).

Fig. 6.3 Wrong illustration

sudden evolution also occurs. This thinking is opposed to that of the "selfish gene[2]" espoused by Richard Dawkins. Gould and Eldredge labeled selfish gene followers "ultra-Darwinists" (reductionists explaining evolution in terms of genes) and referred to themselves as naturalists (focused on the hierarchical structure of the biological system) [2]. Gould criticized the thinking behind "selfish genes" as being linked to genetic supremacy and progressive evolution [1] (thinking behind original form of evolution in Fig. 6.2a).

The misunderstanding that evolution is progress (=optimization) has been widely disseminated. For example, Gould criticized an illustration (Fig. 6.3) still used in evolution textbooks as perpetuating this mistaken way of thinking. Humans have not become the "highest" form of organism or reached the arrival point of evolution, and other organisms are certainly not inferior. Suffice it to say that the most highly evolved species on earth is not the human species, but that of viruses and bacteria. In other words, Gould said that, in terms of evolution, the organism that has flourished most in history is the bacteria.

In this way, the theory of punctuated equilibrium shook up the prevailing theory of neo-Darwinism that had long ruled the field of evolutionary theory around the world. However, discussions in regard to the rightness or wrongness of this continue to this day.

The French geneticist who received the Nobel Prize, Françoise Jacob, stated that "nature is a repairer, not an engineer." The fact is that evolution is not simply a random search. It is a construction device for creating something suboptimal from past components. We should keep in mind that the objective is not optimality but rather robustness.

References

1. Dawkins, R.: A Devil's Chaplain: Reflections on Hope, Lies, Science, and Love. Mariner Books, Boston (2004)
2. Eldredge, N.: Reinventing Darwin: The Great Debate at the High Table of Evolutionary Theory. Wiley, New York (1995)

[2]This is the thinking that genes dexterously behave so as to reproduce successfully and leave as many descendants as possible. With this thinking, we live in order to allow the genes hosted in our own bodies to survive, and we are seen as a vehicle (vehicle=form of transport, medium) for allowing genes to survive. It was discovered that many living phenomena could be explained based on this way of thinking.

Symbol Index

H. Iba, *Evolutionary Approach to Machine Learning and Deep Neural Networks*, https://doi.org/10.1007/978-981-13-0200-8

Author Index

H. Iba, *Evolutionary Approach to Machine Learning and Deep Neural Networks*, https://doi.org/10.1007/978-981-13-0200-8

Subject Index

H. Iba, *Evolutionary Approach to Machine Learning and Deep Neural Networks*, https://doi.org/10.1007/978-981-13-0200-8

Printed in the United States
By Bookmasters